The

GREAT
DIVORCE

BOOKS BY C. S. LEWIS

A Grief Observed
George MacDonald: An Anthology
Mere Christianity
Miracles
The Abolition of Man
The Great Divorce
The Problem of Pain
The Screwtape Letters (with *"Screwtape Proposes a Toast"*)
The Weight of Glory

ALSO AVAILABLE FROM HARPERCOLLINS

The Chronicles of Narnia:
The Magician's Nephew
The Lion, the Witch and the Wardrobe
The Horse and His Boy
Prince Caspian
The Voyage of the Dawn Treader
The Silver Chair
The Last Battle

The GREAT DIVORCE

A DREAM

C. S. Lewis

"No, there is no escape.
There is no heaven with a little of hell in it—
no plan to retain this or that of the devil in our hearts or our pockets. Out Satan must go, every hair and feather."

GEORGE MACDONALD

HarperOne
An Imprint of HarperCollins*Publishers*

HarperOne

HarperCollins books may be purchased for educational, business, or sales promotional use. For information please e-mail the Special Markets Department at SPsales@harpercollins.com.

HarperCollins Web site: http://www.harpercollins.com

HarperCollins®, ▄®, and HarperOne™ are trademarks of HarperCollins Publishers.

FIRST HARPERCOLLINS PAPERBACK EDITION PUBLISHED IN 2001

Library of Congress Cataloging-in-Publication Data
Lewis, C. S. (Clive Staples), 1898–1963.
The great divorce : a dream / C. S. Lewis.
p. cm.
Originally published: Great Britian : G. Bles, 1946
ISBN 978-0-06-065295-1
1. Good and evil. I. Title.
BJ1401.L4 2000
236'.2—dc21   00-049859

23 24 25 26 27 LBC 89 88 87 86 85

To Barbara Wall
Best and most long-suffering of scribes

PREFACE

Blake wrote the Marriage of Heaven and Hell. If I have written of their Divorce, this is not because I think myself a fit antagonist for so great a genius, nor even because I feel at all sure that I know what he meant. But in some sense or other the attempt to make that marriage is perennial. The attempt is based on the belief that reality never presents us with an absolutely unavoidable 'either-or'; that, granted skill and patience and (above all) time enough, some way of embracing both alternatives can always be found; that mere development or adjustment or refinement will somehow turn evil into good without our being called on for a final and total rejection of anything we should like to retain. This belief I take to be a disastrous error. You cannot take all luggage with you on all journeys; on one journey even your right hand and your right eye may be among the things you have to leave behind. We are not living in a world where all roads are radii of a circle and where all, if followed long enough, will therefore draw gradually nearer and finally meet at the centre: rather in a world where every road, after a few miles, forks into two, and each of those into two again, and at each fork you must make a decision. Even on the biological level life is not like a river but like a tree. It does not move towards unity but away from it and the creatures grow further apart as they increase in perfection. Good, as it ripens, becomes continually more different not only from evil but from other good.

I do not think that all who choose wrong roads perish; but their rescue consists in being put back on the right road. A sum can be put right: but only by going back till you find the error and working it afresh from that point, never by simply going on. Evil can be undone, but it cannot 'develop' into good. Time does not heal it. The spell must be unwound, bit by bit, 'with backward mutters of dissevering power'—or else not. It is still 'either-or'. If we insist on keeping Hell (or even Earth) we shall not see Heaven: if we accept Heaven we shall not be able to retain even the smallest and most intimate souvenirs of Hell. I believe, to be sure, that any man who reaches Heaven will find that what he abandoned (even in plucking out his right eye) has not been lost: that the kernel of what he was really seeking even in his most depraved wishes will be there, beyond expectation, waiting for him in 'the High Countries'. In that sense it will be true for those who have completed the journey (and for no others) to say that good is everything and Heaven everywhere. But we, at this end of the road, must not try to anticipate that retrospective vision. If we do, we are likely to embrace the false and disastrous converse and fancy that everything is good and everywhere is Heaven.

But what, you ask, of earth? Earth, I think, will not be found by anyone to be in the end a very distinct place. I think earth, if chosen instead of Heaven, will turn out to have been, all along, only a region in Hell: and earth, if put second to Heaven, to have been from the beginning a part of Heaven itself.

There are only two things more to be said about this small book. Firstly, I must acknowledge my debt to a writer whose name I have forgotten and whom I read several years ago in a highly coloured American magazine of what they call 'Scientifiction'. The unbendable and unbreakable quality of my heavenly matter was suggested to me by him, though he used the fancy for a different and most ingenious purpose. His hero travelled into the past: and there, very properly, found raindrops that would pierce him like bullets and sandwiches that no strength could bite—because, of course, nothing in the past can be altered. I, with less originality but (I hope) equal propriety; have transferred this to the eternal. If the writer of that story ever reads these lines I ask him to accept my grateful acknowledgement. The second thing is this. I beg readers to remember that this is a fantasy. It has of course—or I intended it to have—a moral. But the transmortal conditions are solely an imaginative supposal: they are not even a guess or a speculation at what may actually await us. The last thing I wish is to arouse factual curiosity about the details of the after-world.

C. S. LEWIS

April, 1945

The

GREAT
DIVORCE

I

I seemed to be standing in a busy queue by the side of a long, mean street. Evening was just closing in and it was raining. I had been wandering for hours in similar mean streets, always in the rain and always in evening twilight. Time seemed to have paused on that dismal moment when only a few shops have lit up and it is not yet dark enough for their windows to look cheering. And just as the evening never advanced to night, so my walking had never brought me to the better parts of the town. However far I went I found only dingy lodging houses, small tobacconists, hoardings from which posters hung in rags, windowless warehouses, goods stations without trains, and bookshops of the sort that sell *The Works of Aristotle*. I never met anyone. But for the little crowd at the bus stop, the whole town seemed to be empty. I think that was why I attached myself to the queue.

I had a stroke of luck right away, for just as I took my stand a little waspish woman who would have been ahead of me snapped out at a man who seemed to be with her, 'Very well, then. I won't go at all. So there,' and left the queue. 'Pray don't imagine,' said the man, in a very dignified voice, 'that I care about going in the least. I have only been trying to please *you*, for peace sake. My own feelings are of course a matter of no importance, I quite understand *that*'—and suiting the action to the word he also walked away. 'Come,' thought I, 'that's two places gained.' I was now next to a very short man with a scowl who glanced at me with an expression of extreme disfavour and observed, rather unnecessarily loudly, to the man beyond him, 'This sort of thing really makes one think twice about going at all.' 'What sort of thing?' growled the other, a big beefy person. 'Well,' said the Short Man, 'this is hardly the sort of society I'm used to as a matter of fact.' 'Huh!' said the Big Man: and then added with a glance at me, 'Don't you stand any sauce from *him*, Mister. You're not *afraid* of him, are you?' Then, seeing I made no move, he rounded suddenly on the Short Man and said, 'Not good enough for you, aren't we? Like your lip.' Next moment he had fetched the Short Man one on the side of the face that sent him sprawling into the gutter. 'Let him lay, let him lay,' said the Big Man to no one in particular. 'I'm a plain man

that's what I am and I got to have my rights same as anyone else, see?' As the Short Man showed no disposition to
rejoin the queue and soon began limping away, I closed
up, rather cautiously, behind the Big Man and congratu-
lated myself on having gained yet another step. A moment
later two young people in front of him also left us arm
in arm. They were both so trousered, slender, giggly and
falsetto that I could be sure of the sex of neither, but it
was clear that each for the moment preferred the other
to the chance of a place in the bus. 'We shall never all get
in,' said a female voice with a whine in it from some four
places ahead of me. 'Change places with you for five bob,
lady,' said someone else. I heard the clink of money and
then a scream in the female voice, mixed with roars of
laughter from the rest of the crowd. The cheated woman
leaped out of her place to fly at the man who had bilked
her, but the others immediately closed up and flung her
out . . . So what with one thing and another the queue had
reduced itself to manageable proportions long before the
bus appeared.

It was a wonderful vehicle, blazing with golden light,
heraldically coloured. The Driver himself seemed full of
light and he used only one hand to drive with. The other
he waved before his face as if to fan away the greasy
steam of the rain. A growl went up from the queue as he

came in sight. 'Looks as if he had a good time of it, eh?
. . . Bloody pleased with himself, I bet . . . My dear, why
can't he behave *naturally?*—Thinks himself too good to
look at us . . . Who does he imagine he is? . . . All that
gilding and purple, I call it a wicked waste. Why don't
they spend some of the money on their house prop-
erty down here?—God! I'd like to give him one in the
ear-'ole.' I could see nothing in the countenance of the
Driver to justify all this, unless it were that he had a look
of authority and seemed intent on carrying out his job.

My fellow passengers fought like hens to get on board
the bus though there was plenty of room for us all. I
was the last to get in. The bus was only half full and I
selected a seat at the back, well away from the others. But
a tousle-haired youth at once came and sat down beside
me. As he did so we moved off.

'I thought you wouldn't mind my tacking on to you,'
he said, 'for I've noticed that you feel just as I do about
the present company. Why on earth they insist on coming I can't imagine. They won't like it at all when we
get there, and they'd really be much more comfortable at
home. It's different for you and me.'

'Do they *like* this place?' I asked.

'As much as they'd like anything,' he answered.

'They've got cinemas and fish and chip shops and advertisements and all the sorts of things they want. The appalling lack of any intellectual life doesn't worry *them*. I realised as soon as I got here that there'd been some mistake. I ought to have taken the first bus but I've fooled about trying to wake people up here. I found a few fellows I'd known before and tried to form a little circle, but they all seem to have sunk to the level of their surroundings. Even before we came here I'd had some doubts about a man like Cyril Blellow. I always thought he was working in a false idiom. But he was at least intelligent: one could get some criticism worth hearing from him, even if he was a failure on the creative side. But now he seems to have nothing left but his self-conceit. The last time I tried to read him some of my own stuff . . . but wait a minute, I'd just like you to look at it.'

Realising with a shudder that what he was producing from his pocket was a thick wad of type-written paper, I muttered something about not having my spectacles and exclaimed, 'Hullo! We've left the ground.'

It was true. Several hundred feet below us, already half hidden in the rain and mist, the wet roofs of the town appeared, spreading without a break as far as the eye could reach.

2

I was not left very long at the mercy of the Tousle-Headed Poet, because another passenger interrupted our conversation: but before that happened I had learned a good deal about him. He appeared to be a singularly ill-used man. His parents had never appreciated him and none of the five schools at which he had been educated seemed to have made any provision for a talent and temperament such as his. To make matters worse he had been exactly the sort of boy in whose case the examination system works out with the maximum unfairness and absurdity. It was not until he reached the university that he began to recognise that all these injustices did not come by chance but were the inevitable results of our economic system. Capitalism did not merely enslave the workers, it also vitiated taste and vulgarised intellect: hence our educational system and hence the lack of 'Recognition' for new genius.

This discovery had made him a Communist. But when the war came along and he saw Russia in alliance with the capitalist governments, he had found himself once more isolated and had to become a conscientious objector. The indignities he suffered at this stage of his career had, he confessed, embittered him. He decided he could serve the cause best by going to America: but then America came into the war too. It was at this point that he suddenly saw Sweden as the home of a really new and radical art, but the various oppressors had given him no facilities for going to Sweden. There were money troubles. His father, who had never progressed beyond the most atrocious mental complacency and smugness of the Victorian epoch, was giving him a ludicrously inadequate allowance. And he had been very badly treated by a girl too. He had thought her a really civilised and adult personality, and then she had unexpectedly revealed that she was a mass of bourgeois prejudices and monogamic instincts. Jealousy, possessiveness, was a quality he particularly disliked. She had even shown herself, at the end, to be mean about money. That was the last straw. He had jumped under a train . . .

I gave a start, but he took no notice.

Even then, he continued, ill luck had continued to dog him. He'd been sent to the grey town. But of course it was a mistake. I would find, he assured me, that all the other passengers would be with me on the return journey. But he would not. He was going to stay 'there'. He felt quite certain that he was going where, at last, his finely critical spirit would no longer be outraged by an uncongenial environment—where he would find 'Recognition' and 'Appreciation'. Meanwhile, since I hadn't got my glasses, he would read me the passage about which Cyril Blellow had been so insensitive . . .

It was just then that we were interrupted. One of the quarrels which were perpetually simmering in the bus had boiled over and for a moment there was a stampede. Knives were drawn: pistols were fired: but it all seemed strangely innocuous and when it was over I found myself unharmed, though in a different seat and with a new companion. He was an intelligent-looking man with a rather bulbous nose and a bowler hat. I looked out of the windows. We were now so high that all below us had become featureless. But fields, rivers, or mountains I did not see, and I got the impression that the grey town still filled the whole field of vision.

'It seems the deuce of a town,' I volunteered, 'and that's what I can't understand. The parts of it that I saw were so empty. Was there once a much larger population?'

'Not at all,' said my neighbour. 'The trouble is that they're so quarrelsome. As soon as anyone arrives he settles in some street. Before he's been there twenty-four hours he quarrels with his neighbour. Before the week is over he's quarrelled so badly that he decides to move. Very likely he finds the next street empty because all the people there have quarrelled with *their* neighbours—and moved. If so he settles in. If by any chance the street is full, he goes further. But even if he stays, it makes no odds. He's sure to have another quarrel pretty soon and then he'll move on again. Finally he'll move right out to the edge of the town and build a new house. You see, it's easy here. You've only got to *think* a house and there it is. That's how the town keeps on growing.'

'Leaving more and more empty streets?'

'That's right. And time's sort of odd here. That place where we caught the bus is thousands of miles from the Civic Centre where all the newcomers arrive from earth. All the people you've met were living near the bus stop: but they'd taken centuries—of our time—to get there, by gradual removals.'

'And what about the earlier arrivals? I mean—there must be people who came from Earth to your town even longer ago.'

'That's right. There are. They've been moving on and on. Getting further apart. They're so far off by now that they could never think of coming to the bus stop at all. Astronomical distances. There's a bit of rising ground near where I live and a chap has a telescope. You can see the lights of the inhabited houses, where those old ones live, millions of miles away. Millions of miles from us and from one another. Every now and then they move further still. That's one of the disappointments. I thought you'd meet interesting historical characters. But you don't: they're too far away.'

'Would they get to the bus stop in time, if they ever set out?'

'Well—theoretically. But it'd be a distance of light-years. And they wouldn't want to by now: not those old chaps like Tamberlaine and Genghis Khan, or Julius Caesar, or Henry the Fifth.'

'Wouldn't want to?'

'That's right. The nearest of those old ones is Napoleon. We know that because two chaps made the journey to see him. They'd started long before I came, of course, but I was there when they came back. About fifteen thousand years of our time it took them. We've picked out the house by now. Just a little pin prick of light and nothing else near it for millions of miles.'

'But they got there?'

'That's right. He'd built himself a huge house all in the Empire style—rows of windows flaming with light, though it only shows as a pin prick from where I live.'

'Did they see Napoleon?'

'That's right. They went up and looked through one of the windows. Napoleon was there all right.'

'What was he doing?'

'Walking up and down—up and down all the time—left-right, left-right—never stopping for a moment. The two chaps watched him for about a year and he never rested. And muttering to himself all the time. "It was Soult's fault. It was Ney's fault. It was Josephine's fault. It was the fault of the Russians. It was the fault of the English." Like that all the time. Never stopped for a moment. A little, fat man and he looked kind of tired. But he didn't seem able to stop it.'

From the vibrations I gathered that the bus was still moving, but there was now nothing to be seen from the windows which confirmed this—nothing but grey void above and below.

'Then the town will go on spreading indefinitely?' I said.

'That's right,' said the Intelligent Man. 'Unless someone can do something about it.'

'How do you mean?'

'Well, as a matter of fact, between you and me and the wall, that's my job at the moment. What's the trouble about this place? Not that people are quarrelsome— that's only human nature and was always the same even on Earth. The trouble is they have no Needs. You get everything you want (not very good quality, of course) by just imagining it. That's why it never costs any trouble to move to another street or build another house. In other words, there's no proper economic basis for any community life. If they needed real shops, chaps would have to stay near where the real shops were. If they needed real houses they'd have to stay near where builders were. It's scarcity that enables a society to exist. Well, that's where I come in. I'm not going on this trip for my health. As far as that goes I don't think it would suit me up there. But if I can come back with some real commodities—anything at all that you could really bite or drink or sit on—why, at once you'd get a demand down in our town. I'd start a little business. I'd have something to sell. You'd soon get people coming to live near—centralisation. Two fully-inhabited streets would accommodate the people that are now spread over a million square miles of empty streets. I'd make a nice little profit and be a public benefactor as well.'

'You mean, if they *had* to live together they'd gradually learn to quarrel less?'

'Well, I don't know about that. I daresay they could be kept a bit quieter. You'd have a chance to build up a police force. Knock some kind of discipline into them. Anyway' (here he dropped his voice) 'it'd be *better*, you know. Everyone admits that. Safety in numbers.'

'Safety from what?' I began, but my companion nudged me to be silent. I changed my question.

'But look here,' said I, 'if they can get everything just by imagining it, why would they want any *real* things, as you call them?'

'Eh? Oh well, they'd like houses that really kept out the rain.'

'Their present houses don't?'

'Well, of course not. How could they?'

'What the devil is the use of building them, then?' The Intelligent Man put his head closer to mine. 'Safety again,' he muttered. 'At least, the feeling of safety. It's all right now: but later on . . . you understand.'

'What?' said I, almost involuntarily sinking my own voice to a whisper.

He articulated noiselessly as if expecting that I understood lipreading. I put my ear up close to his mouth. 'Speak up,' I said. 'It will be dark presently,' he mouthed.

'You mean the evening *is* really going to turn into a night in the end?'

He nodded.

'What's that got to do with it?' said I.

'Well . . . no one wants to be out of doors when that happens.'

'Why?'

His reply was so furtive that I had to ask him several times to repeat it. When he had done so, being a little annoyed (as one so often is with whisperers), I replied without remembering to lower my voice.

'Who are "They"?' I asked. 'And what are you afraid they'll do to you? And why should they come out when it's dark? And what protection could an imaginary house give if there was any danger?'

'Here!' shouted the Big Man. 'Who's talking all that stuff? You stop your whispering, you two, if you don't want a hiding, see? Spreading rumours, that's what I call it. You shut your face, Ikey, see?'

'Quite right. Scandalous. Ought to be prosecuted. How did they get on the bus?' growled the passengers.

A fat clean-shaven man who sat on the seat in front of me leaned back and addressed me in a cultured voice.

'Excuse me,' he said, 'but I couldn't help overhearing parts of your conversation. It is astonishing how these primitive superstitions linger on. I beg your pardon? Oh, God bless my soul, that's all it is. There is not a shred of evidence that this twilight is ever going to turn into a night. There has been a revolution of opinion on that in educated circles. I am surprised that you haven't heard of it. All the nightmare fantasies of our ancestors are being swept away. What we now see in this subdued and delicate half-light is the promise of the dawn: the slow turning of a whole nation towards the light. Slow and imperceptible, of course. "And not through Eastern windows only, When daylight comes, comes in the light." And that passion for "real" commodities which our friend speaks of is only materialism, you know. It's retrogressive. Earth-bound! A hankering for matter. But we look on this spiritual city—for with all its faults it is spiritual—as a nursery in which the creative functions of man, now freed from the clogs of matter, begin to try their wings. A sublime thought.'

Hours later there came a change. It began to grow light in the bus. The greyness outside the windows turned from mud-colour to mother of pearl, then to faintest blue, then to a bright blueness that stung the eyes. We seemed to be floating in a pure vacancy. There were no lands, no sun, no stars in sight: only the radiant abyss. I let down the window beside me. Delicious freshness came in for a second, and then—

'What the hell are you doing?' shouted the Intelligent Man, leaning roughly across me and pulling the window sharply up. 'Want us all to catch our death of cold?'

'Hit him a biff,' said the Big Man.

I glanced round the bus. Though the windows were closed, and soon muffed, the bus was full of light. It was cruel light. I shrank from the faces and forms by which I was surrounded. They were all fixed faces, full not of possibilities but impossibilities, some gaunt, some bloated, some glaring with idiotic ferocity, some drowned beyond recovery in dreams; but all, in one way or another, distorted and faded. One had a feeling that they might fall to pieces at any moment if the light grew much stronger. Then—there was a mirror on the end wall of the bus—I caught sight of my own.

And still the light grew.

3

A cliff had loomed up ahead. It sank vertically beneath us so far that I could not see the bottom, and it was dark and smooth. We were mounting all the time. At last the top of the cliff became visible like a thin line of emerald green stretched tight as a fiddle-string. Presently we glided over that top: we were flying above a level, grassy country through which there ran a wide river. We were losing height now: some of the tallest tree tops were only twenty feet below us. Then, suddenly we were at rest. Everyone had jumped up. Curses, taunts, blows, a filth of vituperation, came to my ears as my fellow-passengers struggled to get out. A moment later, and they had all succeeded. I was alone in the bus, and through the open door there came to me in the fresh stillness the singing of a lark.

I got out. The light and coolness that drenched me were like those of summer morning, early morning a minute or two before the sunrise, only that there was a certain difference. I had the sense of being in a larger space, perhaps even a larger *sort* of space, than I had ever known before: as if the sky were further off and the extent of the green plain wider that they could be on this little ball of earth. I had got 'out' in some sense which made the Solar System itself seem an indoor affair. It gave me a feeling of freedom, but also of exposure, possibly of danger, which continued to accompany me through all that followed. It is the impossibility of communicating that feeling, or even of inducing you to remember it as I proceed, which makes me despair of conveying the real quality of what I saw and heard.

At first, of course, my attention was caught by my fellow-passengers, who were still grouped about in the neighbourhood of the omnibus, though beginning, some of them, to walk forward into the landscape with hesitating steps. I gasped when I saw them. Now that they were in the light, they were transparent—fully transparent when they stood between me and it, smudgy and imperfectly opaque when they stood in the shadow of some tree. They were in fact ghosts: man-shaped stains on the brightness of that air. One could attend to them or ignore them at will as you do with the dirt on a window pane. I noticed that the grass did not bend under their feet: even the dew drops were not disturbed.

Then some re-adjustment of the mind or some focussing of my eyes took place, and I saw the whole phenomenon the other way round. The men were as they had always been; as all the men I had known had been perhaps. It was the light, the grass, the trees that were different; made of some different substance, so much solider than things in our country that men were ghosts by comparison. Moved by a sudden thought, I bent down and tried to pluck a daisy which was growing at my feet. The stalk wouldn't break. I tried to twist it, but it wouldn't twist. I tugged till the sweat stood out on my forehead and I had lost most of the skin off my hands. The little flower was hard, not like wood or even like iron, but like diamond. There was a leaf—a young tender beech-leaf, lying in the grass beside it. I tried to pick the leaf up: my heart almost cracked with the effort, and I believe I did just raise it. But I had to let it go at once; it was heavier than a sack of coal. As I stood, recovering my breath with great gasps and looking down at the daisy, I noticed that I could see the grass not only between my feet but *through* them. I also was a phantom. Who will give me words to express the terror of that discovery? 'Golly!' thought I, 'I'm in for it this time.'

'I don't like it! I don't like it,' screamed a voice. 'It gives me the pip!' One of the ghosts had darted past me, back into the bus. She never came out of it again as far as I know.

The others remained, uncertain.

'Hi, Mister,' said the Big Man, addressing the Driver, 'when have we got to be back?'

'You need never come back unless you want to,' he replied. 'Stay as long as you please.' There was an awkward pause.

'This is simply ridiculous,' said a voice in my ear. One of the quieter and more respectable ghosts had sidled up to me. 'There must be some mismanagement,' he continued. 'What's the sense of allowing all that riff-raff to float about here all day? Look at them. They're not enjoying it. They'd be far happier at home. They don't even know what to do.'

'I don't know very well myself,' said I. 'What does one do?'

'Oh me? I shall be met in a moment or two. I'm expected. I'm not bothering about that. But it's rather unpleasant on one's first day to have the whole place crowded out with trippers. Damn it, one's chief object in coming here at all was to avoid them!'

He drifted away from me. And I began to look about. In spite of his reference to a 'crowd', the solitude was so vast that I could hardly notice the knot of phantoms in the foreground. Greenness and light had almost swallowed them up. But very far away I could see what might be either a great bank of cloud or a range of mountains. Sometimes I could make out in it steep forests, far-withdrawing valleys, and even mountain cities perched on inaccessible summits. At other times it became indistinct. The height was so enormous that my waking sight could not have taken in such an object at all. Light brooded on the top of it: slanting down thence it made long shadows behind every tree on the plain. There was no change and no progression as the hours passed. The promise—or the threat—of sunrise rested immovably up there.

Long after that I saw people coming to meet us. Because they were bright I saw them while they were still very distant, and at first I did not know that they were people at all. Mile after mile they drew nearer. The earth shook under their tread as their strong feet sank into the wet turf. A tiny haze and a sweet smell went up where they had crushed the grass and scattered the dew. Some were naked, some robed. But the naked ones did not seem less adorned, and the robes did not disguise in those who wore them the massive grandeur of muscle and the radiant smoothness of flesh. Some were bearded but no one in that company struck me as being of any particular age. One gets glimpses, even in our country, of that which is ageless—heavy thought in the face of an infant, and frolic childhood in that of a very old man. Here it was all like that. They came on steadily. I did not entirely like it. Two of the ghosts screamed and ran for the bus. The rest of us huddled closer to one another.

4

As the solid people came nearer still I noticed that they were moving with order and determination as though each of them had marked his man in our shadowy company. 'There are going to be affecting scenes,' I said to myself. 'Perhaps it would not be right to look on.' With that, I sidled away on some vague pretext of doing a little exploring. A grove of huge cedars to my right seemed attractive and I entered it. Walking proved difficult. The grass, hard as diamonds to my unsubstantial feet, made me feel as if I were walking on wrinkled rock, and I suffered pains like those of the mermaid in Hans Andersen. A bird ran across in front of me and I envied it. It belonged to that country and was as real as the grass. It could bend the stalks and spatter itself with the dew.

Almost at once I was followed by what I have called the Big Man—to speak more accurately, the Big Ghost.

He in his turn was followed by one of the bright people. 'Don't you know me?' he shouted to the Ghost: and I found it impossible not to turn and attend. The face of the solid spirit—he was one of those that wore a robe—made me want to dance, it was so jocund, so established in its youthfulness.

'Well, I'm damned,' said the Ghost. 'I wouldn't have believed it. It's a fair knock-out. It isn't right, Len, you know. What about poor Jack, eh? You look pretty pleased with yourself, but what I say is, What about poor Jack?'

'He is here,' said the other. 'You will meet him soon, if you stay.'

'But you murdered him.'

'Of course I did. It is all right now.'

'All right, is it? All right for you, you mean. But what about the poor chap himself, laying cold and dead?'

'But he isn't. I have told you, you will meet him soon. He sent you his love.'

'What I'd like to understand,' said the Ghost, 'is what you're here for, as pleased as Punch, you, a bloody murderer, while I've been walking the streets down there and living in a place like a pigsty all these years.'

'That is a little hard to understand at first. But it is

all over now. You will be pleased about it presently. Till then there is no need to bother about it.'

'No need to bother about it? Aren't you ashamed of yourself?'

'No. Not as you mean. I do not look at myself. I have given up myself. I had to, you know, after the murder. That was what it did for me. And that was how everything began.'

'Personally,' said the Big Ghost with an emphasis which contradicted the ordinary meaning of the word, 'Personally, I'd have thought you and I ought to be the other way round. That's my personal opinion.'

'Very likely we soon shall be,' said the other. 'If you'll stop thinking about it.'

'Look at me, now,' said the Ghost, slapping its chest (but the slap made no noise). 'I gone straight all my life. I don't say I was a religious man and I don't say I had no faults, far from it. But I done my best all my life, see? I done my best by everyone, that's the sort of chap I was. I never asked for anything that wasn't mine by rights. If I wanted a drink I paid for it and if I took my wages I done my job, see? That's the sort I was and I don't care who knows it.'

'It would be much better not to go on about that now.'

'Who's going on? I'm not arguing. I'm just telling you the sort of chap I was, see? I'm asking for nothing but my rights. You may think you can put me down because you're dressed up like that (which you weren't when you worked under me) and I'm only a poor man. But I got to have my rights same as you, see?'

'Oh no. It's not so bad as that. I haven't got my rights, or I should not be here. You will not get yours either. You'll get something far better. Never fear.'

'That's just what I say. I haven't got my rights. I always done my best and I never done nothing wrong. And what I don't see is why I should be put below a bloody murderer like you.'

'Who knows whether you will be? Only be happy and come with me.'

'What do you keep on arguing for? I'm only telling you the sort of chap I am. I only want my rights. I'm not asking for anybody's bleeding charity.'

'Then do. At once. Ask for the Bleeding Charity. Everything is here for the asking and nothing can be bought.'

'That may do very well for you, I daresay. If they choose to let in a bloody murderer all because he makes a poor mouth at the last moment, that's their look out.

But I don't see myself going in the same boat as you, see? Why should I? I don't want charity. I'm a decent man and if I had my rights I'd have been here long ago and you can tell them I said so.'

The other shook his head. 'You can never do it like that,' he said. 'Your feet will never grow hard enough to walk on our grass that way. You'd be tired out before we got to the mountains. And it isn't exactly true, you know.' Mirth danced in his eyes as he said it.

'What isn't true?' asked the Ghost sulkily.

'You weren't a decent man and you didn't do your best. We none of us were and none of us did. Lord bless you, it doesn't matter. There is no need to go into it all now.'

'You!' gasped the Ghost. '*You* have the face to tell *me* I wasn't a decent chap?'

'Of course. Must I go into all that? I will tell you one thing to begin with. Murdering old Jack wasn't the worst thing I did. That was the work of a moment and I was half mad when I did it. But I murdered you in my heart, deliberately, for years. I used to lie awake at nights thinking what I'd do to you if I ever got the chance. That is why I have been sent to you now: to ask your forgiveness and to be your servant as long as you need one, and longer if it pleases you. I was the worst. But all the men who worked under you felt the same. You made it hard for us, you know. And you made it hard for your wife too and for your children.'

'You mind your own business, young man,' said the Ghost. 'None of your lip, see? Because I'm not taking any impudence from you about my private affairs.'

'There are no private affairs,' said the other.

'And I'll tell you another thing,' said the Ghost. 'You can clear off, see? You're not wanted. I may be only a poor man but I'm not making pals with a murderer, let alone taking lessons from him. Made it hard for you and your like, did I? If I had you back there I'd show you what work is.'

'Come and show me now,' said the other with laughter in his voice, 'It will be joy going to the mountains, but there will be plenty of work.'

'You don't suppose I'd go with you?'

'Don't refuse. You will never get there alone. And I am the one who was sent to you.'

'So that's the trick, is it?' shouted the Ghost, outwardly bitter, and yet I thought there was a kind of triumph in its voice. It had been entreated: it could make a refusal: and this seemed to it a kind of advantage. 'I thought there'd be some damned nonsense. It's all a clique, all a bloody clique. Tell them I'm not coming, see? I'd rather be damned than go along with you. I came here to get my rights, see? Not to go snivelling along on charity tied onto your apron-strings. If they're too fine to have me without you, I'll go home.' It was almost happy now that it could, in a sense, threaten. 'That's what I'll do,' it repeated, 'I'll go home. I didn't come here to be treated like a dog. I'll go home. That's what I'll do. Damn and blast the whole pack of you . . .' In the end, still grumbling, but whimpering also a little as it picked its way over the sharp grasses, it made off.

5

For a moment there was silence under the cedar trees and then—*pad, pad, pad*—it was broken. Two velvet-footed lions came bouncing into the open space, their eyes fixed upon each other, and started playing some solemn romp. Their manes looked as if they had been just dipped in the river whose noise I could hear close at hand, though the tree hid it. Not greatly liking my company, I moved away to find that river, and after passing some thick flowering bushes, I succeeded. The bushes came almost down to the brink. It was as smooth as the Thames but flowed swiftly like a mountain stream: pale green where trees overhung it but so clear that I could count the pebbles at the bottom. Close beside me I saw another of the Bright People in conversation with a ghost. It was that fat ghost with the cultured voice who had addressed me in the bus, and it seemed to be wearing gaiters.

'My dear boy, I'm delighted to see you,' it was saying to the Spirit, who was naked and almost blindingly white. 'I was talking to your poor father the other day and wondering where you were.'

'You didn't bring him?' said the other.

'Well, no. He lives a long way from the bus, and, to be quite frank, he's been getting a little eccentric lately. A little difficult. Losing his grip. He never was prepared to make any great efforts, you know. If you remember, he used to go to sleep when you and I got talking seriously! Ah, Dick, I shall never forget some of our talks. I expect you've changed your views a bit since then. You became rather narrow-minded towards the end of your life: but no doubt you've broadened out again.'

'How do you mean?'

'Well, it's obvious by now, isn't it, that you weren't quite right. Why, my dear boy, you were coming to believe in a literal Heaven and Hell!'

'But wasn't I right?'

'Oh, in a spiritual sense, to be sure. I still believe in them in that way. I am still, my dear boy, looking for the Kingdom. But nothing superstitious or mythological...'

'Excuse me. Where do you imagine you've been?'

'Ah, I see. You mean that the grey town with its continual hope of morning (we must all live by hope, must we not?), with its field for indefinite progress, is, in a sense, Heaven, if only we have eyes to see it? That is a beautiful idea.'

'I didn't mean that at all. Is it possible you don't know where you've been?'

'Now that you mention it, I don't think we ever do give it a name. What do you call it?'

'We call it Hell.'

'There is no need to be profane, my dear boy. I may not be very orthodox, in your sense of that word, but I do feel that these matters ought to be discussed simply, and seriously, and reverently.'

'Discuss Hell *reverently*? I meant what I said. You have been in Hell: though if you don't go back you may call it Purgatory.'

'Go on, my dear boy, go on. That is so like you. No doubt you'll tell me why, on your view, I was sent there. I'm not angry.'

'But don't you know? You went there because you are an apostate.'

'Are you serious, Dick?'

'Perfectly.'

'This is worse than I expected. Do you really think

people are penalised for their honest opinions? Even assuming, for the sake of argument, that those opinions were mistaken.'

'Do you really think there are no sins of intellect?'

'There are indeed, Dick. There is hide-bound prejudice, and intellectual dishonesty, and timidity, and stagnation. But honest opinions fearlessly followed—they are not sins.'

'I know we used to talk that way. I did it too until the end of my life when I became what you call narrow. It all turns on what are honest opinions.'

'Mine certainly were. They were not only honest but heroic. I asserted them fearlessly. When the doctrine of the Resurrection ceased to commend itself to the critical faculties which God had given me, I openly rejected it. I preached my famous sermon. I defied the whole chapter. I took every risk.'

'What risk? What was at all likely to come of it except what actually came—popularity, sales for your books, invitations, and finally a bishopric?'

'Dick, this is unworthy of you. What are you suggesting?'

'Friend, I am not suggesting at all. You see, I know now. Let us be frank. Our opinions were not honestly come by.

We simply found ourselves in contact with a certain current of ideas and plunged into it because it seemed modern and successful. At College, you know, we just started automatically writing the kind of essays that got good marks and saying the kind of things that won applause. When, in our whole lives, did we honestly face, in solitude, the one question on which all turned: whether after all the Supernatural might not in fact occur? When did we put up one moment's real resistance to the loss of our faith?'

'If this is meant to be a sketch of the genesis of liberal theology in general, I reply that it is a mere libel. Do you suggest that men like . . .'

'I have nothing to do with any generality. Nor with any man but you and me. Oh, as you love your own soul, remember. You know that you and I were playing with loaded dice. We didn't want the other to be true. We were afraid of crude salvationism, afraid of a breach with the spirit of the age, afraid of ridicule, afraid (above all) of real spiritual fears and hopes.'

'I'm far from denying that young men may make mistakes. They may well be influenced by current fashions of thought. But it's not a question of how the opinions are formed. The point is that they were my honest opinions, sincerely expressed.'

'Of course. Having allowed oneself to drift, unresist-ing, unpraying, accepting every half-conscious solicita-tion from our desires, we reached a point where we no longer believed the Faith. Just in the same way, a jealous man, drifting and unresisting, reaches a point at which he believes lies about his best friend: a drunkard reaches a point at which (for the moment) he actually believes that another glass will do him no harm. The beliefs are sincere in the sense that they do occur as psychological events in the man's mind. If that's what you mean by sincerity they are sincere, and so were ours. But errors which are sincere in that sense are not innocent.'

'You'll be justifying the Inquisition in a moment!'

'Why? Because the Middle Ages erred in one direc-tion, does it follow that there is no error in the opposite direction?'

'Well, this is extremely interesting,' said the Episcopal Ghost. 'It's a point of view. Certainly, it's a point of view. In the meantime . . .'

'There is no meantime,' replied the other. 'All that is over. We are not playing now. I have been talking of the past (your past and mine) only in order that you may turn from it forever. One wrench and the tooth will be out. You can begin as if nothing had ever gone wrong. White as snow. It's all true, you know. He is in me, for you, with that power. And—I have come a long journey to meet you. You have seen Hell: you are in sight of Heaven. Will you, even now, repent and believe?'

'I'm not sure that I've got the exact point you are trying to make,' said the Ghost.

'I am not trying to make any point,' said the Spirit. 'I am telling you to repent and believe.'

'But my dear boy, I believe already. We may not be perfectly agreed, but you have completely misjudged me if you do not realise that my religion is a very real and a very precious thing to me.'

'Very well,' said the other, as if changing his plan. 'Will you believe in me?'

'In what sense?'

'Will you come with me to the mountains? It will hurt at first, until your feet are hardened. Reality is harsh to the feet of shadows. But will you come?'

'Well, that is a plan. I am perfectly ready to consider it. Of course I should require some assurances . . . I should want a guarantee that you are taking me to a place where I shall find a wider sphere of usefulness—and scope for the talents that God has given me—and an atmosphere of free inquiry—in short, all that one means by civilisation and—er—the spiritual life.'

'No,' said the other. 'I can promise you none of these things. No sphere of usefulness: you are not needed there at all. No scope for your talents: only forgiveness for having perverted them. No atmosphere of inquiry, for I will bring you to the land not of questions but of answers, and you shall see the face of God.'

'Ah, but we must all interpret those beautiful words in our own way! For me there is no such thing as a final answer. The free wind of inquiry must always continue to blow through the mind, must it not? "Prove all things" . . . to travel hopefully is better than to arrive.'

'If that were true, and known to be true, how could anyone travel hopefully? There would be nothing to hope for.'

'But you must feel yourself that there is something stifling about the idea of finality? Stagnation, my dear boy, what is more soul-destroying than stagnation?'

'You think that, because hitherto you have experienced truth only with the abstract intellect. I will bring you where you can taste it like honey and be embraced by it as by a bridegroom. Your thirst shall be quenched.'

'Well, really, you know, I am not aware of a thirst for some ready-made truth which puts an end to intellectual activity in the way you seem to be describing. Will it leave me the free play of Mind, Dick? I must insist on that, you know.'

'Free, as a man is free to drink while he is drinking. He is not free still to be dry.' The Ghost seemed to think for a moment. 'I can make nothing of that idea,' it said.

'Listen!' said the White Spirit. 'Once you were a child. Once you knew what inquiry was for. There was a time when you asked questions because you wanted answers, and were glad when you had found them. Become that child again: even now.'

'Ah, but when I became a man I put away childish things.'

'You have gone far wrong. Thirst was made for water; inquiry for truth. What you now call the free play of inquiry has neither more nor less to do with the ends for which intelligence was given you than masturbation has to do with marriage.'

'If we cannot be reverent, there is at least no need to be obscene. The suggestion that I should return at my age to the mere factual inquisitiveness of boyhood strikes me as preposterous. In any case, that question-and-answer conception of thought only applies to matters of fact. Religious and speculative questions are surely on a different level.'

'We know nothing of religion here: we think only of Christ. We know nothing of speculation. Come and see. I will bring you to Eternal Fact, the Father of all other facthood.'

'I should object very strongly to describing God as a "fact". The Supreme Value would surely be a less inadequate description. It is hardly . . .'

'Do you not even believe that He exists?'

'Exists? What does Existence mean? You *will* keep on implying some sort of static, ready-made reality which is, so to speak, "there", and to which our minds have simply to conform. These great mysteries cannot be approached in that way. If there were such a thing (there is no need to interrupt, my dear boy) quite frankly, I should not be interested in it. It would be of no *religious* significance. God, for me, is something purely spiritual. The spirit of sweetness and light and tolerance—and, er, service, Dick, service. We mustn't forget that, you know.'

'If the thirst of the Reason is really dead . . . ,' said the Spirit, and then stopped as though pondering. Then suddenly he said, 'Can you, at least, still desire happiness?'

'Happiness, my dear Dick,' said the Ghost placidly, 'happiness, as you will come to see when you are older, lies in the path of duty. Which reminds me . . . Bless my soul, I'd nearly forgotten. Of course I can't come with you. I have to be back next Friday to read a paper. We have a little Theological Society down there. Oh yes! there is plenty of intellectual life. Not of a very high quality, perhaps. One notices a certain lack of grip—a certain confusion of mind. That is where I can be of some use to them. There are even regrettable jealousies . . . I don't know why, but tempers seem less controlled than they used to be. Still, one mustn't expect too much of human nature. I feel I can do a great work among them. But you've never asked me what my paper is about! I'm taking the text about growing up to the measure of the stature of Christ and working out an idea which I feel sure you'll be interested in. I'm going to point out how people always forget that Jesus (here the Ghost bowed) was a comparatively young man when he died. He would have outgrown some of his earlier views, you know, if he'd lived. As he might have done, with a little more tact and patience. I am going to ask my audience to consider what his mature views would have been. A profoundly interesting question. What a different Christianity we might have had if only the Founder had reached his full stature! I shall end up by pointing out how this deepens the significance of the Crucifixion. One feels for the first time what a disaster it was: what a tragic waste . . . so much promise cut short. Oh, must you be going? Well, so must I. Goodbye, my dear boy. It has been a great pleasure. Most stimulating and provocative. Goodbye, goodbye, goodbye.'

The Ghost nodded its head and beamed on the Spirit with a bright clerical smile—or with the best approach to it which such unsubstantial lips could manage—and then turned away humming softly to itself 'City of God, how broad and far.'

But I did not watch him for long, for a new idea had just occurred to me. If the grass were hard as rock, I thought, would not the water be hard enough to walk on? I tried it with one foot, and my foot did not go in. Next moment I stepped boldly out on the surface. I fell on my face at once and got some nasty bruises. I had forgotten that though it was, to me, solid, it was not the less in rapid motion. When I had picked myself up I was about thirty yards further down-stream than the point where I had left the bank. But this did not prevent me from walking up-stream: it only meant that by walking very fast indeed I made very little progress.

6

The cool smooth skin of the bright water was delicious to my feet and I walked on it for about an hour, making perhaps a couple of hundred yards. Then the going became difficult. The current grew swifter. Great flakes or islands of foam came swirling down towards me, bruising my shins like stones if I did not get out of their way. The surface became uneven, rounded itself into lovely hollows and elbows of water which distorted the appearance of the pebbles on the bottom and threw me off my balance, so that I had to scramble to shore. But as the banks hereabouts consisted of great flat stones, I continued my journey without much hurt to my feet. An immense yet lovely noise vibrated through the forest. Hours later I rounded a bend and saw the explanation.

Before me green slopes made a wide amphitheatre, enclosing a frothy and pulsating lake into which, over many-coloured rocks, a waterfall was pouring. Here once again I realised that something had happened to my senses so that they were now receiving impressions which would normally exceed their capacity. On Earth, such a waterfall could not have been perceived at all as a whole; it was too big. Its sound would have been a terror in the woods for twenty miles. Here, after the first shock, my sensibility 'took' both as a well-built ship takes a huge wave. I exulted. The noise, though gigantic, was like giants' laughter: like the revelry of a whole college of giants together laughing, dancing, singing, roaring at their high works.

Near the place where the fall plunged into the lake there grew a tree. Wet with the spray, half-veiled in foam-bows, flashing with the bright, innumerable birds that flew among its branches, it rose in many shapes of billowy foliage, huge as a fen-land cloud. From every point apples of gold gleamed through the leaves.

Suddenly my attention was diverted by a curious appearance in the foreground. A hawthorn bush not twenty yards away seemed to be behaving oddly. Then I saw that it was not the bush but something standing close to the bush and on this side of it. Finally I realised that it was one of the Ghosts. It was crouching as if to conceal itself from something beyond the bush, and it was looking back at me and making signals. It kept on signing to me to duck down. As I could not see what the danger was, I stood fast.

Presently the Ghost, after peering around in every direction, ventured beyond the hawthorn bush. It could not get on very fast because of the torturing grasses beneath its feet, but it was obviously going as fast as it possibly could, straight for another tree. There it stopped again, standing straight upright against the trunk as though it were taking cover. Because the shadow of the branches now covered it, I could see it better: it was my bowler-hatted companion, the one whom the Big Ghost had called Ikey. After it had stood panting at the tree for about ten minutes and carefully reconnoitred the ground ahead, it made a dash for another tree—such a dash as was possible to it. In this way, with infinite labour and caution, it had reached the great Tree in about an hour. That is, it had come within ten yards of it.

Here it was checked. Round the Tree grew a belt of lilies: to the Ghost an insuperable obstacle. It might as well have tried to tread down an anti-tank trap as to walk on them. It lay down and tried to crawl between them but they grew too close and they would not bend. And all the time it was apparently haunted by the terror of discovery. At every whisper of the wind it stopped and cowered: once, at the cry of a bird, it struggled back to its last place of cover: but then desire hounded it out again and it crawled once more to the Tree. I saw it clasp its hands and writhe in the agony of its frustration.

The wind seemed to be rising. I saw the Ghost wring its hand and put its thumb into its mouth—cruelly pinched, I doubt not, between two stems of the lilies when the breeze swayed them. Then came a real gust. The branches of the Tree began to toss. A moment later and half a dozen apples had fallen round the Ghost and on it. He gave a sharp cry, but suddenly checked it. I thought the weight of the golden fruit where it had fallen on him would have disabled him: and certainly, for a few minutes, he was unable to rise. He lay whimpering, nursing his wounds. But soon he was at work again. I could see him feverishly trying to fill his pockets with the apples. Of course it was useless. One could see how his ambitions were gradually forced down. He gave up the idea of a pocketful: two would have to do. He gave up the idea of two, he would take one, the largest one. He gave up that hope. He was now looking for the smallest one. He was trying to find if there was one small enough to carry.

The amazing thing was that he succeeded. When I

remembered what the leaf had felt like when I tried to lift it, I could hardly help admiring this unhappy creature when I saw him rise staggering to his feet actually holding the smallest of the apples in his hands. He was lame from his hurts, and the weight bent him double. Yet even so, inch by inch, still availing himself of every scrap of cover, he set out on his *via dolorosa* to the bus, carrying his torture.

'Fool. Put it down,' said a great voice suddenly. It was quite unlike any other voice I had heard so far. It was a thunderous yet liquid voice. With an appalling certainty I knew that the waterfall itself was speaking: and I saw now (though it did not cease to look like a waterfall) that it was also a bright angel who stood, like one crucified, against the rocks and poured himself perpetually down towards the forest with loud joy.

'Fool', he said, 'put it down. You cannot take it back. There is not room for it in Hell. Stay here and learn to eat such apples. The very leaves and the blades of grass in the wood will delight to teach you.'

Whether the Ghost heard or not, I don't know. At any rate, after pausing for a few minutes, it braced itself anew for its agonies and continued with even greater caution till I lost sight of it.

7

Although I watched the misfortunes of the Ghost in the Bowler with some complacency, I found, when we were left alone, that I could not bear the presence of the Water-Giant. It did not appear to take any notice of me, but I became self-conscious; and I rather think there was some assumed nonchalance in my movements as I walked away over the flat rocks, down-stream again. I was beginning to be tired. Looking at the silver fish which darted over the river-bed, I wished greatly that to me also that water were permeable. I should have liked a dip.

'Thinking of going back?' said a voice close at hand. I turned and saw a tall ghost standing with its back against a tree, chewing a ghostly cheroot. It was that of a lean hard-bitten man with grey hair and a gruff, but not uneducated voice: the kind of man I have always instinctively felt to be reliable.

'I don't know,' said I. 'Are you?'

'Yes,' it replied. 'I guess I've seen about all there is to see.'

'You don't think of staying?'

'That's all propaganda,' it said. 'Of course there never was any question of our staying. You can't eat the fruit and you can't drink the water and it takes you all your time to walk on the grass. A human being couldn't live here. All that idea of staying is only an advertisement stunt.'

'Then why did you come?'

'Oh, I don't know. Just to have a look round. I'm the sort of chap who likes to see things for himself. Wherever I've been I've always had a look at anything that was being cracked up. When I was out East, I went to see Pekin. When . . .'

'What was Pekin like?'

'Nothing to it. Just one darn wall inside another. Just a trap for tourists. I've been pretty well everywhere. Niagara Falls, the Pyramids, Salt Lake City, the Taj Mahal . . .'

'What was it like?'

'Not worth looking at. They're all advertisement stunts. All run by the same people. There's a combine, you know, a World Combine, that just takes an Atlas and decides where they'll have a Sight. Doesn't matter what they choose: anything'll do as long as the publicity's properly managed.'

'And you've lived—er—*down there*—in the Town—for some time?'

'In what they call Hell? Yes. It's a flop too. They lead you to expect red fire and devils and all sorts of interesting people sizzling on grids—Henry VIII and all that—but when you get there it's just like any other town.'

'I prefer it up here,' said I.

'Well, I don't see what all the talk is about,' said the Hard-Bitten Ghost. 'It's as good as any other park to look at, and darned uncomfortable.'

'There seems to be some idea that if one stays here one would get—well, solider—grow acclimatised.'

'I know all about that,' said the Ghost. 'Same old lie. People have been telling me that sort of thing all my life. They told me in the nursery that if I were good I'd be happy. And they told me at school that Latin would get easier as I went on. After I'd been married a month some fool was telling me that there were always difficulties at first, but with Tact and Patience I'd soon "settle down" and like it! And all through two wars what didn't they say about the good time coming if only I'd be a brave boy and go on being shot at? Of course they'll play the old game here if anyone's fool enough to listen.'

'But who are "They"? This might be run by someone different?'

'Entirely new management, eh? Don't you believe it! It's never a new management. You'll always find the same old Ring. I know all about dear, kind Mummie coming up to your bedroom and getting all she wants to know out of you: but you always found she and Father were the same firm really. Didn't we find that both sides in all the wars were run by the same Armament Firms? or the same Firm, which is behind the Jews and the Vatican and the Dictators and the Democracies and all the rest of it. All this stuff up here is run by the same people as the Town. They're just laughing at us.'

'I thought they were at war?'

'Of course you did. That's the official version. But who's ever seen any signs of it? Oh, I know that's how they *talk*. But if there's a real war why don't they do anything? Don't you see that if the official version were true these chaps up here would attack and sweep the Town out of existence? They've got the strength. If they wanted to rescue us they could do it. But obviously the last thing they want is to end their so-called "war". The whole game depends on keeping it going.'

This account of the matter struck me as uncomfortably plausible. I said nothing.

'Anyway,' said the Ghost, 'who wants to be rescued? What the hell would there be to *do* here?'

'Or there?' said I.

'Quite,' said the Ghost. 'They've got you either way.'

'What would you like to do if you had your choice?' I asked.

'There you go!' said the Ghost with a certain triumph. 'Asking *me* to make a plan. It's up to the Management to find something that doesn't bore us, isn't it? It's their job. Why should we do it for them? That's just where all the parsons and moralists have got the thing upside down. They keep on asking us to alter ourselves. But if the people who run the show are so clever and so powerful, why don't they find something to suit their public? All this poppycock about growing harder so that the grass doesn't hurt our feet, now! There's an example. What would you say if you went to a hotel where the eggs were all bad and when you complained to the Boss, instead of apologising and changing his dairyman, he just told you that if you tried you'd get to like bad eggs in time?'

'Well, I'll be getting along,' said the Ghost after a short silence. 'You coming my way?'

'There doesn't seem to be much point in going anywhere on your showing,' I replied. A great depression had come over me. 'And at least it's not raining here.'

'Not at the moment,' said the Hard-Bitten Ghost. 'But I never saw one of those bright mornings that didn't turn to rain later on. And, by gum, when it does rain here! Ah, you hadn't thought of that? It hadn't occurred to you that with the sort of water they have here every raindrop will make a hole in you, like a machine-gun bullet. That's their little joke, you see. First of all tantalise you with ground you can't walk on and water you can't drink and then drill you full of holes. But they won't catch me that way.'

A few minutes later he moved off.

8

I sat still on a stone by the river's side feeling as miserable as I ever felt in my life. Hitherto it had not occurred to me to doubt the intentions of the Solid People, nor to question the essential goodness of their country even if it were a country which I could not long inhabit. It had indeed once crossed my mind that if these Solid People were as benevolent as I had heard one or two of them claim to be, they might have done something to help the inhabitants of the Town—something more than meeting them on the plain. Now a terrible explanation came into my mind. How if this whole trip were allowed the Ghosts merely to mock them? Horrible myths and doctrines stirred in my memory. I thought how the Gods had punished Tantalus. I thought of the place in the Book of Revelation where it says that the smoke of Hell goes up forever in the sight of the blessed spirits. I remembered how poor Cowper, dreaming that he was not after all doomed to perdition, at once knew the dream to be false and said, 'These are the sharpest arrows in His quiver.' And what the Hard-Bitten Ghost had said about the rain was clearly true. Even a shower of dew-drops from a branch might tear me in pieces. I had not thought of this before. And how easily I might have ventured into the spray of the waterfall!

The sense of danger, which had never been entirely absent since I left the bus, awoke with sharp urgency. I gazed around on the trees, the flowers, and the talking cataract: they had begun to look unbearably sinister. Bright insects darted to and fro. If one of those were to fly into my face, would it not go right through me? If it settled on my head, would it crush me to earth? Terror whispered, 'This is no place for you.' I remembered also the lions.

With no very clear plan in my mind, I rose and began walking away from the river in the direction where the trees grew closest together. I had not fully made up my mind to go back to the bus, but I wanted to avoid open places. If only I could find a trace of evidence that it was really possible for a Ghost to stay—that the choice was not only a cruel comedy—I would not go back. In the meantime I went on, gingerly, and keeping a sharp lookout. In about half an hour I came to a little clearing with some bushes in the centre. As I stopped, wondering if I dared cross it, I realised that I was not alone.

A Ghost hobbled across the clearing—as quickly as it could on that uneasy soil—looking over its shoulder as if it were pursued. I saw that it had been a woman: a well-dressed woman, I thought, but its shadows of finery looked ghastly in the morning light. It was making for the bushes. It could not really get in among them—the twigs and leaves were too hard—but it pressed as close up against them as it could. It seemed to believe it was hiding.

A moment later I heard the sound of feet, and one of the Bright People came in sight: one always noticed that sound there, for we Ghosts made no noise when we walked.

'Go away!' squealed the Ghost. 'Go away! Can't you see I want to be alone?'

'But you need help,' said the Solid One.

'If you have the least trace of decent feeling left,' said the Ghost, 'you'll keep away. I don't want help. I want to be left alone. Do go away. You know I can't walk fast enough on those horrible spikes to get away from you. It's abominable of you to take advantage.'

'Oh, that!' said the Spirit. 'That'll soon come right. But you're going in the wrong direction. It's back there—to the mountains—you need to go. You can lean on me all the way. I can't absolutely *carry* you, but you need have almost no weight on your own feet: and it will hurt less at every step.'

'I'm not afraid of being hurt. You know that.'

'Then what is the matter?'

'Can't you understand *anything*? Do you really suppose I'm going out there among all those people, like *this*?'

'But why not?'

'I'd never have come at all if I'd known you were all going to be dressed like that.'

'Friend, you see I'm not dressed at all.'

'I didn't mean that. Do go away.'

'But can't you even tell me?'

'If you can't understand, there'd be no good trying to explain it. How can I go out like this among a lot of people with real solid bodies? It's far worse than going out with nothing on would have been on Earth. Have everyone staring *through* me.'

'Oh, I see. But we were all a bit ghostly when we first arrived, you know. That'll wear off. Just come out and try.'

'But they'll see me.'

'What does it matter if they do?'

'I'd rather die.'

'But you've died already. There's no good trying to go back to that.'

The Ghost made a sound something between a sob and a snarl. 'I wish I'd never been born,' it said. 'What *are* we born for?'

'For infinite happiness,' said the Spirit. 'You can step out into it at any moment . . .'

'But, I tell you, they'll *see* me.'

'An hour hence and you will not care. A day hence and you will laugh at it. Don't you remember on earth—there were things too hot to touch with your finger but you could drink them all right? Shame is like that. If you will accept it—if you will drink the cup to the bottom—you will find it very nourishing: but try to do anything else with it and it scalds.'

'You really mean? . . .' said the Ghost, and then paused. My suspense was strained up to the height. I felt that my own destiny hung on her reply. I could have fallen at her feet and begged her to yield.

'Yes,' said the Spirit. 'Come and try.'

Almost, I thought the Ghost had obeyed. Certainly it had moved: but suddenly it cried out, 'No, I can't. I tell you I can't. For a moment, while you were talking, I almost thought . . . but when it comes to the point . . . You've no right to ask me to do a thing like that. It's disgusting. I should never forgive myself if I did. Never, never. And it's not fair. They ought to have warned us. I'd never have come. And now—please, please go away!"

'Friend,' said the Spirit. 'Could you, only for a moment, fix your mind on something not yourself?'

'I've already given you my answer,' said the Ghost, coldly but still tearful.

'Then only one expedient remains,' said the Spirit, and to my great surprise he set a horn to his lips and blew. I put my hands over my ears. The earth seemed to shake: the whole wood trembled and dindled at the sound. I suppose there must have been a pause after that (though there seemed to be none) before I heard the thudding of hoofs—far off at first, but already nearer before I had well identified it, and soon so near that I began to look about for some place of safety. Before I had found one the danger was all about us. A herd of unicorns came thundering through the glades: twenty-seven hands high the smallest of them and white as swans but for the red gleam in eyes and nostrils and the flashing indigo of their horns. I can still remember the squelching noise of the soft wet turf under their hoofs, the breaking of the undergrowth, the snorting and the whinneyings; how their hind legs went up and their horned heads down in mimic battle. Even then I wondered for what real battle it might be the rehearsal. I heard the Ghost scream, and I think it made a bolt away from the bushes . . . perhaps towards the Spirit, but I don't know. For my own nerve failed and I fled, not heeding, for the moment, the horrible going underfoot, and not once daring to pause. So I never saw the end of that interview.

9

'Where are ye going?' said a voice with a strong Scotch accent. I stopped and looked. The sound of the unicorns had long since died away and my flight had brought me to open country. I saw the mountains where the unchanging sunrise lay, and in the foreground two or three pines on a little knoll, with some large smooth rocks, and heather. On one of the rocks sat a very tall man, almost a giant, with a flowing beard. I had not yet looked one of the Solid People in the face. Now, when I did so, I discovered that one sees them with a kind of double vision. Here was an enthroned and shining god, whose ageless spirit weighed upon mine like a burden of solid gold: and yet, at the very same moment, here was an old weather-beaten man, one who might have been a shepherd—such a man as tourists think simple because he is honest and neighbours think 'deep' for the same reason.

His eyes had the far-seeing look of one who has lived long in open, solitary places; and somehow I divined the network of wrinkles which must have surrounded them before re-birth had washed him in immortality.

'I—I don't quite know,' said I.

'Ye can sit and talk to me then,' he said, making room for me on the stone.

'I don't know you, Sir,' said I, taking my seat beside him.

'My name is George,' he answered. 'George MacDonald.'

'Oh!' I cried. 'Then you can tell me! You at least will not deceive me.' Then, supposing that these expressions of confidence needed some explanation, I tried, trembling to tell this man all that his writings had done for me. I tried to tell how a certain frosty afternoon at Leatherhead Station when I first bought a copy of *Phantastes* (being then about sixteen years old) had been to me what the first sight of Beatrice had been to Dante: *Here begins the New Life.* I started to confess how long that Life had delayed in the region of imagination merely: how slowly and reluctantly I had come to admit that his Christendom had more than an accidental connexion with it, how hard I had tried not to see that the true name of the quality which first met me in his books is Holiness. He laid his hand on mine and stopped me.

'Son,' he said, 'Your love—all love—is of inexpressible value to me. But it may save precious time' (here he suddenly looked very Scotch) 'if I inform ye that I am already well acquainted with these biographical details. In fact, I have noticed that your memory misleads you in one or two particulars.'

'Oh!' said I, and became still.

'Ye had started,' said my Teacher, 'to talk of something more profitable.'

'Sir,' said I, 'I had almost forgotten it, and I have no anxiety about the answer now, though I have still a curiosity. It is about these Ghosts. *Do* any of them stay? *Can* they stay? Is any real choice offered to them? How do they come to be here?'

'Did ye never hear of the *Refrigerium?* A man with your advantages might have read of it in Prudentius, not to mention Jeremy Taylor.'

'The name is familiar, Sir, but I'm afraid I've forgotten what it means.'

'It means that the damned have holidays—excursions, ye understand.'

'Excursions to *this* country?'

'For those that will take them. Of course most of the silly creatures don't. They prefer taking trips back to Earth. They go and play tricks on the poor daft women ye call mediums. They go and try to assert their ownership of some house that once belonged to them: and then ye get what's called a Haunting. Or they go to spy on their children. Or literary ghosts hang about public libraries to see if anyone's still reading their books.'

'But if they come here they can really stay?'

'Aye. Ye'll have heard that the emperor Trajan did.'

'But I don't understand. Is judgement not final? Is there really a way out of Hell into Heaven?'

'It depends on the way ye're using the words. If they leave that grey town behind it will not have been Hell. To any that leaves it, it is Purgatory. And perhaps ye had better not call this country Heaven. *Not Deep Heaven*, ye understand.' (Here he smiled at me.) 'Ye can call it the Valley of the Shadow of Life. And yet to those who stay here it will have been Heaven from the first. And ye can call those sad streets in the town yonder the Valley of the Shadow of Death: but to those who remain there they will have been Hell even from the beginning.'

I suppose he saw that I looked puzzled, for presently he spoke again.

'Son,' he said, 'ye cannot in your present state understand eternity: when Anodos looked through the door of
the Timeless he brought no message back. But ye can get
some likeness of it if ye say that both good and evil, when
they are full grown, become retrospective. Not only this
valley but all their earthly past will have been Heaven to
those who are saved. Not only the twilight in that town, but
all their life on Earth too, will then be seen by the damned
to have been Hell. That is what mortals misunderstand.
They say of some temporal suffering, "No future bliss can
make up for it," not knowing that Heaven, once attained,
will work backwards and turn even that agony into a glory.
And of some sinful pleasure they say "Let me have but
this and I'll take the consequences": little dreaming how
damnation will spread back and back into their past and
contaminate the pleasure of the sin. Both processes begin
even before death. The good man's past begins to change so
that his forgiven sins and remembered sorrows take on the
quality of Heaven: the bad man's past already conforms to
his badness and is filled only with dreariness. And that is
why, at the end of all things, when the sun rises here and the
twilight turns to blackness down there, the Blessed will say
"We have never lived anywhere except in Heaven," and the
Lost, "We were always in Hell." And both will speak truly.'

'Is that not very hard, Sir?'

'I mean, that is the real sense of what they will say. In the actual language of the Lost, the words will be different, no doubt. One will say he has always served his country right or wrong; and another that he has sacrificed everything to his Art; and some that they've never been taken in, and some that, thank God, they've always looked after Number One, and nearly all, that, at least they've been true to themselves.'

'And the Saved?'

'Ah, the Saved . . . what happens to them is best described as the opposite of a mirage. What seemed, when they entered it, to be the vale of misery turns out, when they look back, to have been a well; and where present experience saw only salt deserts, memory truthfully records that the pools were full of water.'

'Then those people are right who say that Heaven and Hell are only states of mind?'

'Hush,' he said sternly. 'Do not blaspheme. Hell is a state of mind—ye never said a truer word. And every state of mind, left to itself, every shutting up of the creature within the dungeon of its own mind—is, in the end, Hell. But Heaven is not a state of mind. Heaven is reality itself. All that is fully real is Heavenly. For all that can be shaken will be shaken and only the unshakeable remains.'

'But there is a real choice after death? My Roman Catholic friends would be surprised, for to them souls in Purgatory are already saved. And my Protestant friends would like it no better, for they'd say that the tree lies as it falls.'

'They're both right, maybe. Do not fash yourself with such questions. Ye cannot fully understand the relations of choice and Time till you are beyond both. And ye were not brought here to study such curiosities. What concerns you is the nature of the choice itself: and that ye can watch them making.'

'Well, Sir,' I said, 'That also needs explaining. What do they choose, these souls who go back (I have yet seen no others)? And how can they choose it?'

'Milton was right,' said my Teacher. 'The choice of every lost soul can be expressed in the words "Better to reign in Hell than serve in Heaven." There is always something they insist on keeping even at the price of misery. There is always something they prefer to joy—that is, to reality. Ye see it easily enough in a spoiled child that would sooner miss its play and its supper than say it was sorry and be friends. Ye call it the Sulks. But in adult

life it has a hundred fine names—Achilles' wrath and
Coriolanus' grandeur, Revenge and Injured Merit and
Self-Respect and Tragic Greatness and Proper Pride.'

'Then is no one lost through the undignified vices,
Sir? Through mere sensuality?'

'Some are, no doubt. The sensualist, I'll allow ye,
begins by pursuing a real pleasure, though a small one.
His sin is the less. But the time comes on when, though
the pleasure becomes less and less and the craving fiercer
and fiercer, and though he knows that joy can never
come that way, yet he prefers to joy the mere fondling
of unappeasable lust and would not have it taken from
him. He'd fight to the death to keep it. He'd like well to
be able to scratch; but even when he can scratch no more
he'd rather itch than not.'

He was silent for a few minutes, and then began again.

'Ye'll understand, there are innumerable forms of this
choice. Sometimes forms that one hardly thought of at all
on Earth. There was a creature came here not long ago and
went back—Sir Archibald they called him. In his earthly
life he'd been interested in nothing but Survival. He'd
written a whole shelf-full of books about it. He began by
being philosophical, but in the end he took up Psychical
Research. It grew to be his only occupation—experiment-

ing, lecturing, running a magazine. And travelling too: digging out queer stories among Tibetan lamas and being initiated into brotherhoods in Central Africa. Proofs— and more proofs—and then more proofs again—were what he wanted. It drove him mad if ever he saw anyone taking an interest in anything else. He got into trouble during one of your wars for running up and down the country telling them not to fight because it wasted a lot of money that ought to be spent on Research. Well, in good time, the poor creature died and came here: and there was no power in the universe would have prevented him stay- ing and going on to the mountains. But do ye think that did him any good? This country was no use to him at all. Everyone here had "survived" already. Nobody took the least interest in the question. There was nothing more to prove. His occupation was clean gone. Of course if he would only have admitted that he'd mistaken the means for the end and had a good laugh at himself he could have begun all over again like a little child and entered into joy. But he would not do that. He cared nothing about joy. In the end he went away.'

'How fantastic!' said I.

'Do ye think so?' said the Teacher with a piercing glance. 'It is nearer to such as you than ye think. There have been men before now who got so interested in proving the existence of God that they came to care nothing for God Himself . . . as if the good Lord had nothing to do but exist! There have been some who were so occupied in spreading Christianity that they never gave a thought to Christ. Man! Ye see it in smaller matters. Did ye never know a lover of books that with all his first editions and signed copies had lost the power to read them? Or an organiser of charities that had lost all love for the poor? It is the subtlest of all the snares.'

Moved by a desire to change the subject, I asked why the Solid People, since they were full of love, did not go down into Hell to rescue the Ghosts. Why were they content simply to meet them on the plain? One would have expected a more militant charity.

'Ye will understand that better, perhaps before ye go,' said he. 'In the meantime, I must tell ye they have come further for the sake of the Ghosts than ye can understand. Every one of us lives only to journey further and further into the mountains. Every one of us has interrupted that journey and retraced immeasurable distances to come down today on the mere chance of saving some Ghost. Of course it is also joy to do so, but ye cannot blame us for that! And it would be no use to come further even if it were possible. The sane would do no good if they made themselves mad to help madmen.'

'But what of the poor Ghosts who never get into the omnibus at all?'

'Everyone who wishes it does. Never fear. There are only two kinds of people in the end: those who say to God, "Thy will be done," and those to whom God says, in the end, "Thy will be done." All that are in Hell, choose it. Without that self-choice there could be no Hell. No soul that seriously and constantly desires joy will ever miss it. Those who seek find. To those who knock it is opened.'

At this moment we were suddenly interrupted by the thin voice of a Ghost talking at an enormous speed. Looking behind us we saw the creature. It was addressing one of the Solid People and was doing so too busily to notice us. Every now and then the Solid Spirit tried to get in a word but without success. The Ghost's talk was like this:

'Oh, my dear, I've had such a dreadful time, I don't know how I ever got here at all, I was coming with Elinor Stone and we'd arranged the whole thing and we were to meet at the corner of Sink Street; I made it perfectly plain because I knew what she was like and if I told her once

I told her a hundred times I would not meet her outside that dreadful Marjoribanks woman's house, not after the way she'd treated me . . . that was one of the most dreadful things that happened to me; I've been dying to tell you because I felt sure you'd tell me I acted rightly; no, wait a moment, dear, till I've told you—I tried living with her when I first came and it was all fixed up, she was to do the cooking and I was to look after the house and I *did* think I was going to be comfortable after all I'd been through but she turned out to be so changed, absolutely selfish, and not a particle of sympathy for anyone but herself—and as I once said to her, "I *do* think I'm entitled to a little consideration because you at least lived out your time, but I oughtn't to have been here for years and years yet"—oh but of course I'm forgetting you don't know—I was murdered, simply murdered, dear, that man should never have operated, I ought to be alive today and they simply starved me in that dreadful nursing home and no one ever came near me and . . .'

The shrill monotonous whine died away as the speaker, still accompanied by the bright patience at her side, moved out of hearing.

'What troubles ye, son?' asked my Teacher.

'I am troubled, Sir,' said I, 'because that unhappy creature doesn't seem to me to be the sort of soul that ought to be even in danger of damnation. She isn't wicked: she's only a silly, garrulous old woman who has got into a habit of grumbling, and feels that a little kindness, and rest, and change would do her all right.'

'That is what she once was. That is maybe what she still is. If so, she certainly will be cured. But the whole question is whether she is now a grumbler.'

'I should have thought there was no doubt about that!'

'Aye, but ye misunderstand me. The question is whether she is a grumbler, or only a grumble. If there is a real woman—even the least trace of one—still there inside the grumbling, it can be brought to life again. If there's one wee spark under all those ashes, we'll blow it till the whole pile is red and clear. But if there's nothing but ashes we'll not go on blowing them in our own eyes forever. They must be swept up.'

'But how can there be a grumble without a grumbler?'

'The whole difficulty of understanding Hell is that the thing to be understood is so nearly Nothing. But ye'll have had experiences . . . it begins with a grumbling mood, and yourself still distinct from it: perhaps criticising it. And yourself, in a dark hour, may will that mood, embrace it. Ye can repent and come out of it again. But there may come a day when you can do that no longer. Then there will be no you left to criticise the mood, nor even to enjoy it, but just the grumble itself going on forever like a machine. But come! Ye are here to watch and listen. Lean on my arm and we will go for a little walk.'

I obeyed. To lean on the arm of someone older than myself was an experience that carried me back to childhood, and with this support I found the going tolerable: so much so, indeed, that I flattered myself my feet were already growing more solid, until a glance at the poor transparent shapes convinced me that I owed all this ease to the strong arm of the Teacher. Perhaps it was because of his presence that my other senses also appeared to be quickened. I noticed scents in the air which had hitherto escaped me, and the country put on new beauties. There was water everywhere and tiny flowers quivering in the early breeze. Far off in the woods we saw the deer glancing past, and, once, a sleek panther came purring to my companion's side. We also saw many of the Ghosts.

I think the most pitiable was a female Ghost. Her trouble was the very opposite of that which afflicted the other, the lady frightened by the Unicorns. This one seemed quite unaware of her phantasmal appearance. More than one of the Solid People tried to talk to her, and at first I was quite at a loss to understand her behaviour to them. She appeared to be contorting her all but invisible face and writhing her smokelike body in a quite meaningless fashion. At last I came to the conclusion—incredible as it seemed—that she supposed herself still capable of attracting them and was trying to do so. She was a thing that had become incapable of conceiving conversation save as a means to that end. If a corpse already liquid with decay had arisen from the coffin, smeared its gums with lipstick, and attempted a flirtation, the result could not have been more appalling. In the end she muttered, 'Stupid creatures,' and turned back to the bus.

This put me in mind to ask my Teacher what he thought of the affair with the Unicorns. 'It will maybe have succeeded,' he said. 'Ye will have divined that he meant to frighten her, not that fear itself could make her less a Ghost, but if it took her mind a moment off herself, there might, in that moment, be a chance. I have seen them saved so.'

We met several Ghosts that had come so near to Heaven only in order to tell the Celestials about Hell. Indeed this is one of the commonest types. Others, who had perhaps been (like myself) teachers of some kind actually wanted to give lectures about it: they brought fat notebooks full of statistics, and maps, and (one of them) a magic lantern. Some wanted to tell anecdotes of the notorious sinners of all ages whom they had met below. But the most part seemed to think that the mere fact of having contrived for themselves so much misery gave them a kind of superiority. 'You have led a sheltered life!' they bawled. 'You don't know the seamy side. We'll tell you. We'll give you some hard facts'—as if to tinge Heaven with infernal images and colours had been the only purpose for which they came. All alike, so far as I could judge from my own exploration of the lower world, were wholly unreliable, and all equally incurious about the country in which they had arrived. They repelled every attempt to teach them, and when they found that nobody listened to them they went back, one by one, to the bus.

This curious wish to describe Hell turned out, however, to be only the mildest form of a desire very common among the Ghosts—the desire to *extend* Hell, to bring it bodily, if they could, into Heaven. There were tub-thumping Ghosts who in thin, bat-like voices urged the blessed spirits to shake off their fetters, to escape from their imprisonment in happiness, to tear down the mountains with their hands, to seize Heaven 'for their own': Hell offered her co-operation. There were planning

Ghosts who implored them to dam the river, cut down the trees, kill the animals, build a mountain railway, smooth out the horrible grass and moss and heather with asphalt. There were materialistic Ghosts who informed the immortals that they were deluded: there was no life after death, and this whole country was a hallucination. There were Ghosts, plain and simple: mere bogies, fully conscious of their own decay, who had accepted the traditional role of the spectre, and seemed to hope they could frighten someone. I had had no idea that this desire was possible. But my Teacher reminded me that the pleasure of frightening is by no means unknown on Earth, and also of Tacitus' saying: 'They terrify lest they should fear.' When the debris of a decayed human soul finds itself crumbled into ghosthood and realises 'I myself am now that which all humanity has feared, I am just that cold churchyard shadow, that horrible thing which cannot be, yet somehow is', then to terrify others appears to it an escape from the doom of being a Ghost yet still fearing Ghosts—fearing even the Ghost it is. For to be afraid of oneself is the last horror.

But, beyond all these, I saw other grotesque phantoms in which hardly a trace of the human form remained; monsters who had faced the journey to the bus stop— perhaps for them it was thousands of miles—and come up to the country of the Shadow of Life and limped far into it over the torturing grass, only to Spit and gibber out in one ecstasy of hatred their envy and (what is harder to understand) their contempt, of joy. The voyage seemed to them a small price to pay if once, only once, within sight of that eternal dawn, they could tell the prigs, the toffs, the sanctimonious humbugs, the snobs, the 'haves', what they thought of them.

'How do they come to be here at all?' I asked my Teacher.

'I have seen that kind converted,' said he, 'when those ye would think less deeply damned have gone back. Those that hate goodness are sometimes nearer than those that know nothing at all about it and think they have it already.'

'Whisht, now!' said my Teacher suddenly. We were standing close to some bushes and beyond them I saw one of the Solid People and a Ghost who had apparently just that moment met. The outlines of the Ghost looked vaguely familiar, but I soon realized that what I had seen on Earth was not the man himself but photographs of him in the papers. He had been a famous artist.

'God' said the Ghost, glancing round the landscape.

'God what?' asked the Spirit.

'What do you mean, "God what"?' asked the Ghost.

'In our grammar God is a noun.'

'Oh—I see. I only meant "By Gum" or something of the sort. I meant . . . well, all *this*. It's . . . it's . . . I should like to paint this.'

'I shouldn't bother about that just at present if I were you.'

'Look here; isn't one going to be allowed to go on painting?'

'Looking comes first.'

'But I've had my look. I've seen just what I want to do. God!—I wish I'd thought of bringing my things with me!'

The Spirit shook his head, scattering light from his hair as he did so. 'That sort of thing's no good here,' he said.

'What do you mean?' said the Ghost.

'When you painted on earth—at least in your earlier days—it was because you caught glimpses of Heaven in the earthly landscape. The success of your painting was that it enabled others to see the glimpses too. But here you are having the thing itself. It is from here that the messages came. There is no good *telling* us about this country, for we see it already. In fact we see it better than you do.'

'Then there's never going to be any point in painting here?'

'I don't say that. When you've grown into a Person (it's all right, we all had to do it) there'll be some things which you'll see better than anyone else. One of the things you'll want to do will be to tell us about them. But not yet. At present your business is to see. Come and see. He is endless. Come and feed.'

There was a little pause. 'That will be delightful,' said the Ghost presently in a rather dull voice.

'Come, then,' said the Spirit, offering it his arm.

'How soon do you think I could begin painting?' it asked.

The Spirit broke into laughter. 'Don't you see you'll never paint at all if that's what you're thinking about?' he said.

'What do you mean?' asked the Ghost.

'Why, if you are interested in the country only for the sake of painting it, you'll never learn to see the country.'

'But that's just how a real artist is interested in the country.'

'No. You're forgetting,' said the Spirit. 'That was not how you began. Light itself was your first love: you loved paint only as a means of telling about light.'

'Oh, that's ages ago,' said the Ghost. 'One grows out of that. Of course, you haven't seen my later works. One becomes more and more interested in paint for its own sake.'

'One does, indeed. I also have had to recover from that. It was all a snare. Ink and catgut and paint were necessary down there, but they are also dangerous stimulants. Every poet and musician and artist, but for Grace, is drawn away from love of the thing he tells, to love of the telling till, down in Deep Hell, they cannot be interested in God at all but only in what they say about Him. For it doesn't stop at being interested in paint, you know. They sink lower— become interested in their own personalities and then in nothing but their own reputations.'

'I don't think I'm much troubled in *that* way,' said the Ghost stiffly.

'That's excellent,' said the Spirit. 'Not many of us had quite got over it when we first arrived. But if there is any of that inflammation left it will be cured when you come to the fountain.'

'What fountain's that?'

'It is up there in the mountains,' said the Spirit. 'Very cold and clear, between two green hills. A little like Lethe. When you have drunk of it you forget forever all proprietorship in your own works. You enjoy them just as if they were someone else's: without pride and without modesty.'

'That'll be grand,' said the Ghost without enthusiasm.

'Well, come,' said the Spirit: and for a few paces he supported the hobbling shadow forward to the East.

'Of course,' said the Ghost, as if speaking to itself, 'there'll always be interesting people to meet . . .'

'Everyone will be interesting.'

'Oh—ah—yes, to be sure. I was thinking of people in our own line. Shall I meet Claude? Or Cézanne? Or—.'

'Sooner or later—if they're here.'

'But don't you know?'

'Well, of course not. I've only been here a few years. All the chances are against my having run across them . . . there are a good many of us, you know.'

'But surely in the case of distinguished people, you'd hear?'

'But they aren't distinguished—no more than anyone else. Don't you understand? The Glory flows into everyone, and back from everyone: like light and mirrors. But the light's the thing.'

'Do you mean there are no famous men?'

'They are all famous. They are all known, remembered, recognised by the only Mind that can give a perfect judgement.'

'Oh, of course, in *that* sense . . .' said the Ghost.

'Don't stop,' said the Spirit, making to lead him still forward.

'One must be content with one's reputation among posterity, then,' said the Ghost.

'My friend,' said the Spirit. 'Don't you know?'

'Know what?'

'That you and I are already completely forgotten on the Earth?'

'Eh? What's that?' exclaimed the Ghost, disengaging its arm. 'Do you mean those damned Neo-Regionalists have won after all?'

'Lord love you, yes!' said the Spirit, once more shaking and shining with laughter. 'You couldn't get five pounds for any picture of mine or even of yours in Europe or America to-day. We're dead out of fashion.'

'I must be off at once,' said the Ghost. 'Let me go! Damn it all, one has one's duty to the future of Art. I must go back to my friends. I must write an article. There must be a manifesto. We must start a periodical. We must have publicity. Let me go. This is beyond a joke!'

And without listening to the Spirit's reply, the spectre vanished.

10

This conversation also we overheard.

'That is quite, *quite* out of the question,' said a female Ghost to one of the bright Women, 'I should not dream of staying if I'm expected to meet Robert. I am ready to forgive him, of course. But anything more is quite impossible. How he comes to be here . . . but that is your affair.'

'But if you have forgiven him,' said the other, 'surely —.'

'I forgive him as a Christian,' said the Ghost. 'But there are some things one can never forget.'

'But I don't understand . . .' began the She-Spirit.

'Exactly,' said the Ghost with a little laugh. 'You never did. You always thought Robert could do no wrong. *I* know. Please don't interrupt for *one* moment. You haven't the faintest conception of what I went through with your dear Robert. The ingratitude! It was I who made a man of him! Sacrificed my whole life to him! And what was my reward? Absolute, utter selfishness. No, but listen. He was pottering along on about six hundred a year when I married him. And mark my words, Hilda, he'd have been in that position to the day of his death if it hadn't been for me. It was I who had to drive him every step of the way. He hadn't a spark of ambition. It was like trying to lift a sack of coal. I had to positively nag him to take on that extra work in the other department, though it was really the beginning of everything for him. The laziness of men! He said, if you please, he couldn't work more than thirteen hours a day! As if I weren't working far longer. For *my* day's work wasn't over when his was. I had to keep him going all evening, if you understand what I mean. If he'd had his way he'd have just sat in an armchair and sulked when dinner was over. It was I who had to draw him out of himself and brighten him up and make conversation. With no help from him, of course. Sometimes he didn't even listen. As I said to him, I should have thought good manners, if nothing else . . . he seemed to have forgotten that I was a lady even if I *had* married him, and all the time I was working my fingers to the bone for him: and without the slightest appreciation. I used to spend simply *hours* arranging flowers to make that poky little house nice, and instead of thanking me, what do you think he said? He said he wished I wouldn't fill up the writing desk with them when he wanted to use it: and there was a perfectly frightful fuss one evening because I'd spilled one of the vases over some papers of his. If was all nonsense really, because they weren't anything to do with his work. He had some silly idea of writing a book in those days . . . as if he could. I cured him of that in the end.

'No, Hilda, you *must* listen to me. The trouble I went to, entertaining! Robert's idea was that he'd just slink off by himself every now and then to see what he called his old friends . . . and leave me to amuse myself! But I knew from the first that those friends were doing him no good. "No, Robert," said I, "your friends are now mine. It is my duty to have them *here,* however tired I am and however little we can afford it." You'd have thought that would have been enough. But they did come for a bit. That is where I had to use a certain amount of tact. A woman who has her wits about her can always drop in a word here and there. I wanted Robert to see them against a different background. They weren't quite at their ease, somehow, in my drawing-room: not at their best. I couldn't help laughing sometimes. Of course Robert was uncomfortable while the treatment was going on, but it was all for his own good in the end. None of that set were friends of his any longer by the end of the first year.

'And then, he got the new job. A great step up. But what do you think? Instead of realising that we now had a chance to spread out a bit, all he said was "Well *now*, for God's sake let's have some peace." That nearly finished me. I nearly gave him up altogether: but I knew my duty. I have always done my duty. You can't believe the work I had getting him to agree to a bigger house, and then finding a house. I wouldn't have grudged it one scrap if only he'd taken it in the right spirit—if only he'd seen the *fun* of it all. If he'd been a different sort of man it *would* have been fun meeting him on the doorstep as he came back from the office and saying, "Come along, Bobs, no time for dinner to-night. I've just heard of a house near Watford and I've got the keys and we can get there and back by one o'clock." But with *him!* It was perfect misery, Hilda. For by this time your wonderful Robert was turning into the sort of man who cares about nothing but food.

'Well, I got him into the new house at last. Yes, I know. It was a little more than we could really afford at the moment, but all sorts of things were opening out before him. And, of course, I began to entertain properly. No more of his sort of friends, thank you. I was doing it all for his sake. Every useful friend he ever made was due to me. Naturally, I had to dress well. They ought to have been the happiest years of both our lives. If they weren't, he had no one but himself to thank. Oh, he was a maddening man, simply maddening! He just set himself to get old and silent and grumpy. Just sank into himself. He could have looked years younger if he'd taken the trouble. He needn't have walked with a stoop—I'm sure I warned him about that often enough. He was the most miserable host. Whenever we gave a party everything rested on my shoulders: Robert was simply a wet blanket. As I said to him (and if I said it once, I said it a hundred times) he hadn't always been like that. There had been a time when he took an interest in all sorts of things and had been quite ready to make friends. "What on earth is coming over you?" I used to say. But now he just didn't answer at all. He would sit staring at me with his great big eyes (I came to hate a man with dark eyes) and—I know it now—just hating me. That was my reward. After all I'd done. Sheer wicked, senseless hatred: at the very moment when he was a richer man that he'd ever dreamed of being! As I used to say to him, "Robert, you're simply letting yourself go to seed." The younger men who came to the house—it wasn't my fault if they liked me better than my old bear of a husband— used to laugh at him.

'I did my duty to the very end. I *forced* him to take exercise—that was really my chief reason for keeping a great Dane. I kept on giving parties. I took him for the most wonderful holidays. I saw that he didn't drink too much. Even, when things became desperate, I encouraged him to take up his writing again. It couldn't do any harm by then. How could I help it if he *did* have a nervous breakdown in the end? My conscience is clear. I've done my duty by him, if ever a woman has. So you see why it would be impossible to . . .

'And yet . . . I don't know. I believe I have changed my mind. I'll make them a fair offer, Hilda. I will not meet him, if it means just meeting him and no more. But if I'm given a free hand I'll take charge of him again. I will take up my burden once more. But I must have a free hand. With all the time one would have here, I believe I could still make something of him. Somewhere quiet to ourselves. Wouldn't that be a good plan? He's not fit to be on his own. Put me in charge of him. He wants firm handling. I know him better than you do. What's that? No, give him to me, do you hear? Don't consult *him:*

just give him to me. I'm his wife, aren't I? I was only beginning. There's lots, lots, lots of things I still want to do with him. No, listen, Hilda. Please, please! I'm so miserable. I must have someone to—to do things to. It's simply frightful down there. No one minds about me at all. I can't alter them. It's dreadful to see them all sitting about and not be able to do anything with them. Give him back to me. Why should he have everything his own way? It's not good for him. It isn't right, it's not fair. I want Robert. What right have you to keep him from me? I hate you. How can I pay him out if you won't let me have him?'

The Ghost which had towered up like a dying candle-flame snapped suddenly. A sour, dry smell lingered in the air for a moment and then there was no Ghost to be seen.

I I

One of the most painful meetings we witnessed was between a woman's Ghost and a Bright Spirit who had apparently been her brother. They must have met only a moment before we ran across them, for the Ghost was just saying in a tone of unconcealed disappointment, 'Oh . . . Reginald! It's *you*, is it?'

'Yes, dear,' said the Spirit. 'I know you expected someone else. Can you . . . I hope you can be a little glad to see even me; for the present.'

'I did think Michael would have come,' said the Ghost; and then, almost fiercely, 'He is *here*, of course?'

'He's there—far up in the mountains.'

'Why hasn't he come to meet me? Didn't he know?'

'My dear (don't worry, it will all come right presently) it wouldn't have done. Not yet. He wouldn't be able to see or hear you as you are at present. You'd be totally invisible to Michael. But we'll soon build you up.'

'I should have thought if you can see me, my own son could!'

'It doesn't always happen like that. You see, I have specialised in this sort of work.'

'Oh, it's work, is it?' snapped the Ghost. Then, after a pause, 'Well. When am I going to be allowed to see him?'

'There's no question of being *allowed*, Pam. As soon as it's possible for him to see you, of course he will. You need to be thickened up a bit.'

'How?' said the Ghost. The monosyllable was hard and a little threatening.

'I'm afraid the first step is a hard one,' said the Spirit. 'But after that you'll go on like a house on fire. You will become solid enough for Michael to perceive you when you learn to want Someone Else besides Michael. I don't say "more than Michael", not as a beginning. That will come later. It's only the little germ of a desire for God that we need to start the process.'

'Oh, you mean religion and all that sort of thing? This is hardly the moment . . . and from *you*, of all people. Well, never mind. I'll do whatever's necessary. What do you want me to do? Come on. The sooner I begin it, the sooner they'll let me see my boy. I'm quite ready.'

'But, Pam, do think! Don't you see you are not beginning at all as long as you are in that state of mind? You're treating God only as a means to Michael. But the whole thickening treatment consists in learning to want God for His own sake.'

'You wouldn't talk like that if you were a mother.'

'You mean, if I were *only* a mother. But there is no such thing as being only a mother. You exist as Michael's mother only because you first exist as God's creature. That relation is older and closer. No, listen, Pam! He also loves. He also has suffered. He also has waited a long time.'

'If He loved me He'd let me see my boy. If He loved me why did He take Michael away from me? I wasn't going to say anything about that. But it's pretty hard to forgive, you know.'

'But He had to take Michael away. Partly for Michael's sake . . .'

'I'm sure I did my best to make Michael happy. I gave up my whole life . . .'

'Human beings can't make one another really happy for long. And secondly, for your sake. He wanted your merely instinctive love for your child (tigresses share that, you know!) to turn into something better. He wanted you to love Michael as He understands love. You cannot love a fellow-creature fully till you love God. Sometimes this conversion can be done while the instinctive love is still gratified. But there was, it seems, no chance of that in your case. The instinct was uncontrolled and fierce and monomaniac. (Ask your daughter, or your husband. Ask our own mother. You haven't once thought of *her*.) The only remedy was to take away its object. It was a case for surgery. When that first kind of love was thwarted, then there was just a chance that in the loneliness, in the silence, something else might begin to grow.'

'This is all nonsense—cruel and wicked nonsense. What *right* have you to say things like that about Mother-love? It is the highest and holiest feeling in human nature.'

'Pam, Pam—no natural feelings are high or low, holy or unholy, in themselves. They are all holy when God's hand is on the rein. They all go bad when they set up on their own and make themselves into false gods.'

'My love for Michael would never have gone bad. Not if we'd lived together for millions of years.'

'You are mistaken. And you must know. Haven't you

met—down there—mothers who have their sons with them, in Hell? Does *their* love make them happy?'

'If you mean people like the Guthrie woman and her dreadful Bobby, of course not. I hope you're not suggesting . . . If I had Michael I'd be perfectly happy, even in that town. I wouldn't be always talking about him till everyone hated the sound of his name, which is what Winifred Guthrie does about *her* brat. I wouldn't quarrel with people for not taking enough notice of him and then be furiously jealous if they did. I wouldn't go about whining and complaining that he wasn't nice to me. Because, of course, he would be nice. Don't you dare to suggest that Michael could ever become like the Guthrie boy. There are some things I won't stand.'

'What you have seen in the Guthries is what natural affection turns to in the end if it will not be converted.'

'It's a lie. A wicked, cruel lie. How could anyone love their son more than I did? Haven't I lived only for his memory all these years?'

'That was rather a mistake, Pam. In your heart of hearts you know it was.'

'What was a mistake?'

'All that ten years' ritual of grief. Keeping his room exactly as he'd left it; keeping anniversaries; refusing to leave that house though Dick and Muriel were both wretched there.'

'Of course they didn't care. I know that. I soon learned to expect no real sympathy from them.'

'You're wrong. No man ever felt his son's death more than Dick. Not many girls loved their brothers better than Muriel. It wasn't against Michael they revolted: it was against you—against having their whole life dominated by the tyranny of the past: and not really even Michael's past, but your past.'

'You are heartless. Everyone is heartless. The past was all I had.'

'It was all you chose to have. It was the wrong way to deal with a sorrow. It was Egyptian—like embalming a dead body.'

'Oh, of course. I'm wrong. Everything I say or do is wrong, according to you.'

'But of course!' said the Spirit, shining with love and mirth so that my eyes were dazzled. 'That's what we all find when we reach this country. We've all been wrong! That's the great joke. There's no need to go on pretending one was right! After that we begin living.'

'How dare you laugh about it? Give me my boy. Do you hear? I don't care about all your rules and regulations.

I don't believe in a God who keeps mother and son apart. I believe in a God of love. No one had a right to come between me and my son. Not even God. Tell Him that to His face. I want my boy, and I mean to have him. He is mine, do you understand? Mine, mine, mine, for ever and ever.'

'He will be, Pam. Everything will be yours. God Himself will be yours. But not that way. Nothing can be yours by nature.'

'What? Not my own son, born out of my own body?'

'And where is your own body now? Didn't you know that Nature draws to an end? Look! The sun is coming, over the mountains there: it will be up any moment now.'

'Michael is mine.'

'How yours? You didn't make him. Nature made him to grow in your body without your will. Even against your will . . . you sometimes forget that you didn't intend to have a baby then at all. Michael was originally an Accident.'

'Who told you that?' said the Ghost: and then, recovering itself, 'It's a lie. It's not true. And it's no business of yours. I hate your religion and I hate and despise your God. I believe in a God of Love.'

'And yet, Pam, you have no love at this moment for your own mother or for me.'

'Oh, I see! *That's* the trouble, is it? *Really,* Reginald! The idea of your being hurt because . . .'

'Lord love you!' said the Spirit with a great laugh. 'You needn't bother about that! Don't you know that you *can't* hurt anyone in this country?'

The Ghost was silent and open-mouthed for a moment; more wilted, I thought, by this re-assurance than by anything else that had been said.

'Come. We will go a bit further,' said my Teacher, laying his hand on my arm.

'Why did you bring me away, Sir?' said I when we had passed out of earshot of this unhappy Ghost.

'It might take a long while, that conversation,' said my Teacher. 'And ye have heard enough to see what the choice is.'

'Is there any hope for her, Sir?'

'Aye, there's some. What she calls her love for her son has turned into a poor, prickly, astringent sort of thing. But there's still a wee spark of something that's not just herself in it. That might be blown into a flame.'

'Then some natural feelings are really better than others—I mean, are a better starting-point for the real thing?'

'Better *and* worse. There's something in natural affection which will lead it on to eternal love more easily than natural appetite could be led on. But there's also something in it which makes it easier to stop at the natural level and mistake it for the heavenly. Brass is mistaken for gold more easily than clay is. And if it finally refuses conversion its corruption will be worse than the corruption of what ye call the lower passions. It is a stronger angel, and therefore, when it falls, a fiercer devil.'

'I don't know that I dare repeat this on Earth, Sir,' said I. 'They'd say I was inhuman: they'd say I believed in total depravity: they'd say I was attacking the best and the holiest things. They'd call me . . .'

'It might do you no harm if they did,' said he with (I really thought) a twinkle in his eye.

'But could one dare—could one have the face—to go to a bereaved mother, in her misery—when one's not bereaved oneself? . . .'

'No, no, Son, that's no office of yours. You're not a good enough man for that. When your own heart's been broken it will be time for you to think of talking. But someone must say in general what's been unsaid among you this many a year: that love, as mortals understand the word, isn't enough. Every natural love will rise again and live forever in this country: but none will rise again until it has been buried.'

'The saying is almost too hard for us.'

'Ah, but it's cruel not to say it. They that know have grown afraid to speak. That is why sorrows that used to purify now only fester.'

'Keats was wrong, then, when he said he was certain of the holiness of the heart's affections.'

'I doubt if he knew clearly what he meant. But you and I must be clear. There is but one good; that is God. Everything else is good when it looks to Him and bad when it turns from Him. And the higher and mightier it is in the natural order, the more demoniac it will be if it rebels. It's not out of bad mice or bad fleas you make demons, but out of bad archangels. The false religion of lust is baser than the false religion of mother-love or patriotism or art: but lust is less likely to be made into a religion. But look!'

I saw coming towards us a Ghost who carried something on his shoulder. Like all the Ghosts, he was unsubstantial, but they differed from one another as smokes differ. Some had been whitish; this one was dark and oily. What sat on his shoulder was a little red lizard, and it was twitching its tail like a whip and whispering things in his ear. As we caught sight of him he turned his head to the reptile with a snarl of impatience. 'Shut up, I tell you!' he said. It wagged its tail and continued to whisper to him. He ceased snarling, and presently began to smile. Then he turned and started to limp westward, away from the mountains.

'Off so soon?' said a voice.

The speaker was more or less human in shape but larger than a man, and so bright that I could hardly look at him. His presence smote on my eyes and on my body too (for there was heat coming from him as well as light) like the morning sun at the beginning of a tyrannous summer day.

'Yes. I'm off,' said the Ghost. 'Thanks for all your hospitality. But it's no good, you see. I told this little chap' (here he indicated the Lizard) 'that he'd have to be quiet if he came—which he insisted on doing. Of course his stuff won't do here: I realise that. But he won't stop. I shall just have to go home.'

'Would you like me to make him quiet?' said the flaming Spirit—an angel, as I now understood.

'Of course I would,' said the Ghost.

'Then I will kill him,' said the Angel, taking a step forward.

'Oh—ah—look out! You're burning me. Keep away,' said the Ghost, retreating.

'Don't you *want* him killed?'

'You didn't say anything about *killing* him at first. I hardly meant to bother you with anything so drastic as that.'

'It's the only way,' said the Angel, whose burning hands were now very close to the Lizard. 'Shall I kill it?'

'Well, that's a further question. I'm quite open to consider it, but it's a new point, isn't it? I mean, for the moment I was only thinking about silencing it because up here—well, it's so damned embarrassing.'

'May I kill it?'

'Well, there's time to discuss that later.'

'There is no time. May I kill it?'

'Please, I never meant to be such a nuisance. Please—really—don't bother. Look! It's gone to sleep of its own accord. I'm sure it'll be all right now. Thanks ever so much.'

'May I kill it?'

'Honestly, I don't think there's the slightest necessity for that. I'm sure I shall be able to keep it in order now. I think the gradual process would be far better than kill-ing it.'

'The gradual process is of no use at all.'

'Don't you think so? Well, I'll think over what you've said very carefully. I honestly will. In fact I'd let you kill it now, but as a matter of fact I'm not feeling frightfully well to-day. It would be most silly to do it now. I'd need to be in good health for the operation. Some other day, perhaps.'

'There is no other day. All days are present now.'

'Get back! You're burning me. How can I tell you to kill it? You'd kill *me* if you did.'

'It is not so.'

'Why, you're hurting me now.'

'I never said it wouldn't hurt you. I said it wouldn't kill you.'

'Oh, I know. You think I'm a coward. But it isn't that. Really it isn't. I say! Let me run back by to-night's bus and get an opinion from my own doctor. I'll come again the first moment I can.'

'This moment contains all moments.'

'Why are you torturing me? You are jeering at me. How can I let you tear me in pieces? If you wanted to help me, why didn't you kill the damned thing without asking me—before I knew? It would be all over by now if you had.'

'I cannot kill it against your will. It is impossible. Have I your permission?'

The Angel's hands were almost closed on the Lizard,

but not quite. Then the Lizard began chattering to the Ghost so loud that even I could hear what it was saying.

'Be careful,' it said. 'He can do what he says. He can kill me. One fatal word from you and he *will!* Then you'll be without me for ever and ever. It's not natural. How could you live? You'd be only a sort of ghost, not a real man as you are now. He doesn't understand. He's only a cold, bloodless abstract thing. It may be natural for him, but it isn't for us. Yes, yes. I know there are no real pleasures now, only dreams. But aren't they better than nothing? And I'll be so good. I admit I've some-times gone too far in the past, but I promise I won't do it again. I'll give you nothing but really nice dreams—all sweet and fresh and almost innocent. You might say, quite innocent . . .'

'Have I your permission?' said the Angel to the Ghost.

'I know it will kill me.'

'It won't. But supposing it did?'

'You're right. It would be better to be dead than to live with this creature.'

'Then I may?'

'Damn and blast you! Go on, can't you? Get it over. Do what you like,' bellowed the Ghost: but ended, whimpering, 'God help me. God help me.'

Next moment the Ghost gave a scream of agony such as I never heard on Earth. The Burning One closed his crimson grip on the reptile: twisted it, while it bit and writhed, and then flung it, broken-backed, on the turf.

'Ow! That's done for me,' gasped the Ghost, reeling backwards.

For a moment I could make out nothing distinctly. Then I saw, between me and the nearest bush, unmistakably solid but growing every moment solider, the upper arm and the shoulder of a man. Then, brighter still and stronger, the legs and hands. The neck and golden head materialised while I watched, and if my attention had not wavered I should have seen the actual completing of a man—an immense man, naked, not much smaller than the Angel. What distracted me was the fact that at the same moment something seemed to be happening to the Lizard. At first I thought the operation had failed. So far from dying, the creature was still struggling and even growing bigger as it struggled. And as it grew it changed. Its hinder parts grew rounder. The tail, still flickering, became a tail of hair that flickered between huge and glossy buttocks. Suddenly I started back, rubbing my eyes. What stood before me was the greatest stallion I have ever seen, silvery white but with mane and tail of gold. It was smooth and

shining, rippled with swells of flesh and muscle, whinney-
ing and stamping with its hoofs. At each stamp the land
shook and the trees dindled.

The new-made man turned and clapped the new
horse's neck. It nosed his bright body. Horse and master
breathed each into the other's nostrils. The man turned
from it, flung himself at the feet of the Burning One, and
embraced them. When he rose I thought his face shone
with tears, but it may have been only the liquid love and
brightness (one cannot distinguish them in that country)
which flowed from him. I had not long to think about it.
In joyous haste the young man leaped upon the horse's
back. Turning in his seat he waved a farewell, then nudged
the stallion with his heels. They were off before I knew
well what was happening. There was riding if you like!
I came out as quickly as I could from among the bushes
to follow them with my eyes; but already they were only
like a shooting star far off on the green plain, and soon
among the foothills of the mountains. Then, still like a
star, I saw them winding up, scaling what seemed impossible steeps, and quicker every moment, till near the dim
brow of the landscape, so high that I must strain my neck
to see them, they vanished, bright themselves, into the
rose-brightness of that everlasting morning.

While I still watched, I noticed that the whole plain and forest were shaking with a sound which in our world would be too large to hear, but there I could take it with joy. I knew it was not the Solid People who were singing. It was the voice of that earth, those woods and those waters. A strange archaic, inorganic noise, that came from all directions at once. The Nature or Arch-Nature of that land rejoiced to have been once more ridden, and therefore consummated, in the person of the horse. It sang,

'The Master says to our master, Come up. Share my rest and splendour till all natures that were your enemies become slaves to dance before you and backs for you to ride, and firmness for your feet to rest on.

'From beyond all place and time, out of the very Place, authority will be given you: the strengths that once opposed your will shall be obedient fire in your blood and heavenly thunder in your voice.

'Overcome us that, so overcome, we may be ourselves: we desire the beginning of your reign as we desire dawn and dew, wetness at the birth of light.

'Master, your Master has appointed you for ever: to be our King of Justice and our high Priest.'

'Do ye understand all this, my Son?' said the Teacher.

'I don't know about *all*, Sir,' said I. 'Am I right in thinking the Lizard really turned into the Horse?'

'Aye. But it was killed first. Ye'll not forget that part of the story?'

'I'll try not to, Sir. But does it mean that everything—everything—that is in us can go on to the Mountains?'

'Nothing, not even the best and noblest, can go on as it now is. Nothing, not even what is lowest and most bestial, will not be raised again if it submits to death. It is sown a natural body, it is raised a spiritual body. Flesh and blood cannot come to the Mountains. Not because they are too rank, but because they are too weak. What is a lizard compared with a stallion? Lust is a poor, weak, whimpering, whispering thing compared with that richness and energy of desire which will arise when lust has been killed.'

'But am I to tell them at home that this man's sensuality proved less of an obstacle than that poor woman's love for her son? For that was, at any rate, an excess of love.'

'Ye'll tell them no such thing,' he replied sternly. 'Excess of love, did ye say? There was no excess, there was defect. She loved her son too little, not too much. If she had loved him more there'd be no difficulty. I do not know how her affair will end. But it may well be that at this moment she's demanding to have him down with her in Hell. That kind is sometimes perfectly ready to plunge the soul they say they love in endless misery if only they can still in some fashion possess it. No, no. Ye must draw another lesson. Ye must ask, if the risen body even of appetite is as grand a horse as ye saw, what would the risen body of maternal love or friendship be?'

But once more my attention was diverted. 'Is there *another* river, Sir?' I asked.

12

The reason why I asked if there were another river was this. All down one long aisle of the forest the under-sides of the leafy branches had begun to tremble with dancing light; and on Earth I knew nothing so likely to produce this appearance as the reflected lights cast upward by moving water. A few moments later I realised my mistake. Some kind of procession was approaching us, and the light came from the persons who composed it.

First came bright Spirits, not the Spirits of men, who danced and scattered flowers—soundlessly falling, lightly drifting flowers, though by the standards of the ghost-world each petal would have weighed a hundred-weight and their fall would have been like the crashing of boulders. Then, on the left and right, at each side of the forest avenue, came youthful shapes, boys upon one hand, and girls upon the other. If I could remember their singing and write down the notes, no man who read that score would ever grow sick or old. Between them went musicians: and after these a lady in whose honour all this was being done.

I cannot now remember whether she was naked or clothed. If she were naked, then it must have been the almost visible penumbra of her courtesy and joy which produces in my memory the illusion of a great and shining train that followed her across the happy grass. If she were clothed, then the illusion of nakedness is doubtless due to the clarity with which her innermost spirit shone through the clothes. For clothes in that country are not a disguise: the spiritual body lives along each thread and turns them into living organs. A robe or a crown is there as much one of the wearer's features as a lip or an eye.

But I have forgotten. And only partly do I remember the unbearable beauty of her face.

'Is it? . . . is it?' I whispered to my guide.

'Not at all,' said he. 'It's someone ye'll never have heard of. Her name on Earth was Sarah Smith and she lived at Golders Green.'

'She seems to be . . . well, a person of particular importance?'

'Aye. She is one of the great ones. Ye have heard that

fame in this country and fame on Earth are two quite different things.'

'And who are these gigantic people . . . look! They're like emeralds . . . who are dancing and throwing flowers before her?'

'Haven't ye read your Milton? *A thousand liveried angels lackey her.*'

'And who are all these young men and women on each side?'

'They are her sons and daughters.'

'She must have had a very large family, Sir.'

'Every young man or boy that met her became her son—even if it was only the boy that brought the meat to her back door. Every girl that met her was her daughter.'

'Isn't that a bit hard on their own parents?'

'No. There are those that steal other people's children. But her motherhood was of a different kind. Those on whom it fell went back to their natural parents loving them more. Few men looked on her without becoming, in a certain fashion, her lovers. But it was the kind of love that made them not less true, but truer, to their own wives.'

'And how . . . but hullo! What are all these animals? A cat—two cats—dozens of cats. And all these dogs . . . why, I can't count them. And the birds. And the horses.'

'They are her beasts.'

'Did she keep a sort of zoo? I mean, this is a bit too much.'

'Every beast and bird that came near her had its place in her love. In her they became themselves. And now the abundance of life she has in Christ from the Father flows over into them.'

I looked at my Teacher in amazement.

'Yes,' he said. 'It is like when you throw a stone into a pool, and the concentric waves spread out further and further. Who knows where it will end? Redeemed humanity is still young, it has hardly come to its full strength. But already there is joy enough in the little finger of a great saint such as yonder lady to waken all the dead things of the universe into life.'

While we spoke the Lady was steadily advancing towards us, but it was not at us she looked. Following the direction of her eyes, I turned and saw an oddly-shaped phantom approaching. Or rather two phantoms: a great tall Ghost, horribly thin and shaky, who seemed to be leading on a chain another Ghost no bigger than an organ-grinder's monkey. The taller Ghost wore a soft black hat, and he reminded me of something that my memory could not quite recover. Then, when he had come within a few feet of the Lady he spread out his lean, shaky hand flat on his chest with the fingers wide apart, and exclaimed in a hollow voice, 'At last!' All at once I realised what it was that he had put me in mind of. He was like a seedy actor of the old school.

'Darling! At last!' said the Lady. 'Good Heavens!' thought I. 'Surely she can't —', and then I noticed two things. In the first place, I noticed that the little Ghost was not being led by the big one. It was the dwarfish figure that held the chain in its hand and the theatrical figure that wore the collar round its neck. In the second place, I noticed that the Lady was looking solely at the dwarf Ghost. She seemed to think it was the Dwarf who had addressed her, or else she was deliberately ignoring the other. On the poor dwarf she turned her eyes. Love shone not from her face only, but from all her limbs, as if it were some liquid in which she had just been bathing. Then, to my dismay, she came nearer. She stooped down and kissed the Dwarf. It made one shudder to see her in such close contact with that cold, damp, shrunken thing. But she did not shudder.

'Frank,' she said, 'before anything else, forgive me. For all I ever did wrong and for all I did not do right since the first day we met, I ask your pardon.'

I looked properly at the Dwarf for the first time now: or perhaps, when he received her kiss he became a little more visible. One could just make out the sort of face he must have had when he was a man: a little, oval, freckled face with a weak chin and a tiny wisp of unsuccessful moustache. He gave her a glance, not a full look. He was watching the Tragedian out of the corner of his eyes. Then he gave a jerk to the chain: and it was the Tragedian, not he, who answered the Lady.

'There, there,' said the Tragedian. 'We'll say no more about it. We all make mistakes.' With the words there came over his features a ghastly contortion which, I think, was meant for an indulgently playful smile. 'We'll say no more,' he continued. 'It's not myself I'm thinking about. It is you. That is what has been continually on my mind—all these years. The thought of you—you here alone, breaking your heart about me.'

'But now,' said the Lady to the Dwarf, 'you can set all that aside. Never think like that again. It is all over.'

Her beauty brightened so that I could hardly see anything else, and under that sweet compulsion the Dwarf really looked at her for the first time. For a second I thought he was growing more like a man. He opened his mouth. He himself was going to speak this time. But oh, the disappointment when the words came!

'You missed me?' he croaked in a small, bleating voice.

Yet even then she was not taken aback. Still the love and courtesy flowed from her.

'Dear, you will understand about that very soon,' she said. 'But to-day —.'

What happened next gave me a shock. The Dwarf and Tragedian spoke in unison, not to her but to one another. 'You'll notice,' they warned one another, 'she hasn't answered our question.' I realised then that they were one person, or rather that both were the remains of what had once been a person. The Dwarf again rattled the chain.

'You missed me?' said the Tragedian to the Lady, throwing a dreadful theatrical tremor into his voice.

'Dear friend,' said the Lady, still attending exclusively to the Dwarf, 'you may be happy about that and about everything else. Forget all about it for ever.'

And really, for a moment, I thought the Dwarf was going to obey: partly because the outlines of his face became a little clearer, and partly because the invitation to all joy, singing out of her whole being like a bird's song on an April evening, seemed to me such that no creature could resist it. Then he hesitated. And then—once more he and his accomplice spoke in unison.

'Of course it would be rather fine and magnanimous not to press the point,' they said to one another. 'But can we be sure she'd notice? We've done these sort of things before. There was the time we let her have the last stamp in the house to write to her mother and said nothing although she *had* known we wanted to write a letter ourself. We'd thought she'd remember and see how unselfish we'd been. But she never did. And there was the time . . . oh, lots and lots of times!' So the Dwarf gave a shake to the chain and —

'I can't forget it,' cried the Tragedian. 'And I won't forget it, either. I could forgive them all they've done to me. But for your miseries —.'

'Oh, don't you understand?' said the Lady. 'There *are* no miseries here.'

'Do you mean to say,' answered the Dwarf, as if this new idea had made him quite forget the Tragedian for a moment, 'do you mean to say you've been *happy?*'

'Didn't you want me to be? But no matter. Want it now. Or don't think about it at all.'

The Dwarf blinked at her. One could see an unheard-of idea trying to enter his little mind: one could see even that there was for him some sweetness in it. For a second he had almost let the chain go: then, as if it were his lifeline, he clutched it once more.

'Look here,' said the Tragedian. 'We've got to face this.' He was using his 'manly' bullying tone this time: the one for bringing women to their senses.

'Darling,' said the Lady to the Dwarf, 'there's nothing to face. You don't want me to have been miserable for misery's sake. You only think I must have been if I loved you. But if you'll only wait you'll see that isn't so.'

'Love!' said the Tragedian striking his forehead with his hand: then, a few notes deeper, 'Love! Do you know the meaning of the word?'

'How should I not?' said the Lady. 'I am in love. *In* love, do you understand? Yes, now I love truly.'

'You mean,' said the Tragedian, 'you mean—you did not love me truly in the old days.'

'Only in a poor sort of way,' she answered. 'I have asked you to forgive me. There was a little real love in it. But what we called love down there was mostly the craving to be loved. In the main I loved you for my own sake: because I needed you.'

'And now!' said the Tragedian with a hackneyed gesture of despair. 'Now, you need me no more?'

'But of course not!' said the Lady; and her smile made me wonder how both the phantoms could refrain from crying out with joy.

'What needs could I have,' she said, 'now that I have all? I am full now, not empty. I am in Love Himself, not lonely. Strong, not weak. You shall be the same. Come and see. We shall have no *need* for one another now: we can begin to love truly.'

But the Tragedian was still striking attitudes. 'She needs me no more—no more. No more,' he said in a choking voice to no one in particular. 'Would to God,' he continued, but he was now pronouncing it *Gud*— 'would to Gud I had seen her lying dead at my feet before I heard those words. Lying dead at my feet. Lying dead at my feet.'

I do not know how long the creature intended to go on repeating the phrase, for the Lady put an end to that. 'Frank! Frank!' she cried in a voice that made the whole wood ring. '*Look* at me. Look at me. What are you doing with that great, ugly doll? Let go of the chain. Send it away. It is *you* I want. Don't you see what nonsense it's talking.' Merriment danced in her eyes. She was sharing a joke with the Dwarf, right over the head of the Tragedian. Something not at all unlike a smile struggled to appear on the Dwarf's face. For he was looking at her now. Her laughter was past his first defences. He was struggling hard to keep it out, but already with imperfect success. Against his will, he was even growing a little bigger. 'Oh, you great goose,' said she. 'What is the good of talking like that here? You know as well as I do that you *did* see me lying dead years and years ago. Not "at your feet", of course, but on a bed in a nursing home. A very good nursing home it was too. Matron would never have dreamed of leaving bodies lying about the floor! It's ridiculous for that doll to try to be impressive about death *here.* It just won't work.'

13

I do not know that I ever saw anything more terrible than the struggle of that Dwarf Ghost against joy. For he had almost been overcome. Somewhere, incalculable ages ago, there must have been gleams of humour and reason in him. For one moment, while she looked at him in her love and mirth, he saw the absurdity of the Tragedian. For one moment he did not at all misunderstand her laughter: he too must once have known that no people find each other more absurd than lovers. But the light that reached him, reached him against his will. This was not the meeting he had pictured; he would not accept it. Once more he clutched at his death-line, and at once the Tragedian spoke.

'You dare to laugh at it!' it stormed. 'To my face? And this is my reward. Very well. It is fortunate that you give yourself no concern about my fate. Otherwise you might be sorry afterwards to think that you had driven me back to Hell. What? Do you think I'd stay *now*? Thank you. I believe I'm fairly quick at recognising where I'm not wanted. "Not needed" was the exact expression, if I remember rightly.'

From this time on the Dwarf never spoke again: but still the Lady addressed it.

'Dear, no one sends you back. Here is all joy. Everything bids you stay.' But the Dwarf was growing smaller even while she spoke.

'Yes,' said the Tragedian. 'On terms you might offer to a dog. I happen to have some self-respect left, and I see that my going will make no difference to you. It is nothing to you that I go back to the cold and the gloom, the lonely, lonely streets —.'

'Don't, don't, Frank,' said the Lady. 'Don't let it talk like that.' But the Dwarf was now so small that she had dropped on her knees to speak to it. The Tragedian caught her words greedily as a dog catches a bone.

'Ah, you can't bear to hear it!' he shouted with miserable triumph. 'That was always the way. *You* must be sheltered. Grim realities must be kept out of *your* sight. You who can be happy without me, forgetting me! You don't want even to hear of my sufferings. You say, *don't.*

Don't tell you. Don't make you unhappy. Don't break in on your sheltered, self-centred little heaven. And this is the reward —.'

She stooped still lower to speak to the Dwarf which was now a figure no bigger than a kitten, hanging on the end of the chain with his feet off the ground.

'That wasn't why I said, Don't,' she answered. 'I meant, stop acting. It's no good. He is killing you. Let go of that chain. Even now.'

'Acting,' screamed the Tragedian. 'What do you mean?'

The Dwarf was now so small that I could not distinguish him from the chain to which he was clinging. And now for the first time I could not be certain whether the Lady was addressing him or the Tragedian.

'Quick,' she said. 'There is still time. Stop it. Stop it at once.'

'Stop what?'

'Using pity, other people's pity, in the wrong way. We have all done it a bit on earth, you know. Pity was meant to be a spur that drives joy to help misery. But it can be used the wrong way round. It can be used for a kind of blackmailing. Those who choose misery can hold joy up to ransom, by pity. You see, I know now. Even as a child you did it. Instead of saying you were sorry, you went and sulked in the attic . . . because you knew that sooner or later one of your sisters would say, "I can't bear to think of him sitting up there alone, crying." You used their pity to blackmail them, and they gave in in the end. And afterwards, when we were married . . . oh, it doesn't matter, if only you will *stop* it.'

'And *that*,' said the Tragedian, 'that is all you have understood of me, after all these years.' I don't know what had become of the Dwarf Ghost by now. Perhaps it was climbing up the chain like an insect: perhaps it was somehow absorbed into the chain.

'No, Frank, not *here*,' said the Lady. 'Listen to reason. Did you think joy was created to live always under that threat? Always defenceless against those who would rather be miserable than have their self-will crossed? For it was real misery. I know that now. You made yourself really wretched. That you can still do. But you can no longer communicate your wretchedness. Everything becomes more and more itself. Here is joy that cannot be shaken. Our light can swallow up your darkness: but your darkness cannot now infect our light. No, no, no. Come to us. We will not go to you. Can you really have thought that love and joy would always be at the mercy of frowns and sighs? Did you not know they were stronger than their opposites?'

'Love? How dare you use that sacred word?' said the Tragedian. At the same moment he gathered up the chain which had now for some time been swinging uselessly at his side, and somehow disposed of it. I am not quite sure, but I think he swallowed it. Then for the first time it became clear that the Lady saw and addressed him only.

'Where is Frank?' she said. 'And who are you, Sir? I never knew you. Perhaps you had better leave me. Or stay, if you prefer. If it would help you and if it were possible I would go down with you into Hell: but you cannot bring Hell into me.'

'You do not love me,' said the Tragedian in a thin bat-like voice: and he was now very difficult to see.

'I cannot love a lie,' said the Lady. 'I cannot love the thing which is not. I am in Love, and out of it I will not go.'

There was no answer. The Tragedian had vanished. The Lady was alone in that woodland place, and a brown bird went hopping past her, bending with its light feet the grasses I could not bend.

Presently the lady got up and began to walk away. The other Bright Spirits came forward to receive her, singing as they came:

'The Happy Trinity is her home: nothing can trouble her
joy.

She is the bird that evades every net: the wild deer that
leaps every pitfall.

Like the mother bird to its chickens or a shield to the
arm'd knight: so is the Lord to her mind, in His
unchanging lucidity.

Bogies will not scare her in the dark: bullets will not
frighten her in the day.

Falsehoods tricked out as truths assail her in vain: she
sees through the lie as if it were glass.

The invisible germ will not harm her: nor yet the
glittering sun-stroke.

A thousand fail to solve the problem, ten thousand choose
the wrong turning: but she passes safely through.

He details immortal gods to attend her: upon every road
where she must travel.

They take her hand at hard places: she will not stub her
toes in the dark.

She may walk among Lions and rattlesnakes: among
dinosaurs and nurseries of lionets.

He fills her brim-full with immensity of life: he leads her
to see the world's desire.'

'And yet . . . and yet . . . ,' said I to my Teacher, when all the shapes and the singing had passed some distance away into the forest, 'even now I am not quite sure. Is it really tolerable that she should be untouched by his misery, even his self-made misery?'

'Would ye rather he still had the power of tormenting her? He did it many a day and many a year in their earthly life.'

'Well, no. I suppose I don't want that.'

'What then?'

'I hardly know, Sir. What some people say on Earth is that the final loss of one soul gives the lie to all the joy of those who are saved.'

'Ye see it does not.'

'I feel in a way that it ought to.'

'That sounds very merciful: but see what lurks behind it.'

'What?'

'The demand of the loveless and the self-imprisoned that they should be allowed to blackmail the universe: that till they consent to be happy (on their own terms) no one else shall taste joy: that theirs should be the final power; that Hell should be able to *veto* Heaven.'

'I don't know what I want, Sir.'

'Son, son, it must be one way or the other. Either the day must come when joy prevails and all the makers of misery are no longer able to infect it: or else for ever and ever the makers of misery can destroy in others the happiness they reject for themselves. I know it has a grand sound to say ye'll accept no salvation which leaves even one creature in the dark outside. But watch that sophistry or ye'll make a Dog in a Manger the tyrant of the universe.'

'But dare one say—it is horrible to say—that Pity must ever die?'

'Ye must distinguish. The action of Pity will live for ever: but the passion of Pity will not. The passion of Pity, the Pity we merely suffer, the ache that draws men to concede what should not be conceded and to flatter when they should speak truth, the pity that has cheated many a woman out of her virginity and many a statesman out of his honesty—that will die. It was used as a weapon by bad men against good ones: their weapon will be broken.'

'And what is the other kind—the action?'

'It's a weapon on the other side. It leaps quicker than light from the highest place to the lowest to bring healing and joy, whatever the cost to itself. It changes darkness into light and evil into good. But it will not, at the cunning tears of Hell, impose on good the tyranny of evil. Every disease that submits to a cure shall be cured: but we will not call blue yellow to please those who insist on still having jaundice, nor make a midden of the world's garden for the sake of some who cannot abide the smell of roses.'

'You say it will go down to the lowest, Sir. But she didn't go down with him to Hell. She didn't even see him off by the bus.'

'Where would ye have had her go?'

'Why, where we all came from by that bus. The big gulf, beyond the edge of the cliff. Over there. You can't see it from here, but you must know the place I mean.'

My Teacher gave a curious smile. 'Look,' he said, and with the word he went down on his hands and knees. I did the same (how it hurt my knees!) and presently saw that he had plucked a blade of grass. Using its thin end as a pointer, he made me see, after I had looked very closely, a crack in the soil so small that I could not have identified it without this aid.

'I cannot be certain,' he said, 'that this is the crack ye came up through. But through a crack no bigger than that ye certainly came.'

'But—but,' I gasped with a feeling of bewilderment not unlike terror. 'I saw an infinite abyss. And cliffs towering up and up. And then *this* country on top of the cliffs.'

'Aye. But the voyage was not mere locomotion. That bus, and all you inside it, were increasing *in size*.'[1]

'Do you mean then that Hell—all that infinite empty town—is down in some little crack like this?'

'Yes. All Hell is smaller than one pebble of your earthly world: but it is smaller than one atom of *this* world, the Real World. Look at yon butterfly. If it swallowed all Hell, Hell would not be big enough to do it any harm or to have any taste.'

'It seems big enough when you're in it, Sir.'

'And yet all loneliness, angers, hatreds, envies and itchings that it contains, if rolled into one single experience and put into the scale against the least moment of the joy that is felt by the least in Heaven, would have no weight that could be registered at all. Bad cannot succeed even in being bad as truly as good is good. If all Hell's miseries together entered the consciousness of yon wee yellow bird on the bough there, they would be swallowed up without trace, as if one drop of ink had been

---

1. This method of travel also I learned from the 'Scientifictionists'.

dropped into that Great Ocean to which your terrestrial Pacific itself is only a molecule.'

'I see,' said I at last. 'She couldn't *fit* into Hell.'

He nodded. 'There's not room for her,' he said. 'Hell could not open its mouth wide enough.'

'And she couldn't make herself smaller?—like Alice, you know.'

'Nothing like small enough. For a damned soul is nearly nothing: it is shrunk, shut up in itself. Good beats upon the damned incessantly as sound waves beat on the ears of the deaf, but they cannot receive it. Their fists are clenched, their teeth are clenched, their eyes fast shut. First they will not, in the end they cannot, open their hands for gifts, or their mouth for food, or their eyes to see.'

'Then no one can ever reach them?'

'Only the Greatest of all can make Himself small enough to enter Hell. For the higher a thing is, the lower it can descend—a man can sympathise with a horse but a horse cannot sympathise with a rat. Only One has descended into Hell.'

'And will He ever do so again?'

'It was not once long ago that He did it. Time does not work that way when once ye have left the Earth. All moments that have been or shall be were, or are, present in the moment of His descending. There is no spirit in prison to Whom He did not preach.'

'And some hear him?'

'Aye.'

'In your own books, Sir,' said I, 'you were a Universalist. You talked as if all men would be saved. And St. Paul too.'

'Ye can know nothing of the end of all things, or nothing expressible in those terms. It may be, as the Lord said to the Lady Julian, that all will be well, and all will be well, and all manner of things will be well. But it's ill talking of such questions.'

'Because they are too terrible, Sir?'

'No. Because all answers deceive. If ye put the question from within Time and are asking about possibilities, the answer is certain. The choice of ways is before you. Neither is closed. Any man may choose eternal death. Those who choose it will have it. But if ye are trying to leap on into eternity, if ye are trying to see the final state of all things as it will be (for so ye must speak) when there are no more possibilities left but only the Real, then ye ask what cannot be answered to mortal ears. Time is the very lens through which ye see—small and clear, as men see through the wrong end of a telescope—something that would otherwise be too big for ye to see at all. That thing is Freedom: the gift whereby ye most resemble your Maker and are yourselves parts of eternal reality. But ye can see it only through the lens of Time, in a little clear picture, through the inverted telescope. It is a picture of moments following one another and yourself in each moment making some choice that might have been otherwise. Neither the temporal succession nor the phantom of what ye might have chosen and didn't is itself Freedom. They are a lens. The picture is a symbol: but it's truer than any philosophical theorem (or, perhaps, than any mystic's vision) that claims to go behind it. For every attempt to see the shape of eternity except through the lens of Time destroys your knowledge of Freedom. Witness the doctrine of Predestination which shows (truly enough) that eternal reality is not waiting for a future in which to be real; but at the price of removing Freedom which is the deeper truth of the two. And wouldn't Universalism do the same? Ye cannot know eternal reality by a definition. Time itself, and all acts and events that fill Time, are the definition, and it must be lived. The Lord said we were gods. How long could ye bear to look (without Time's lens) on the greatness of your own soul and the eternal reality of her choice?'

14

And suddenly all was changed. I saw a great assembly of gigantic forms all motionless, all in deepest silence, standing forever about a little silver table and looking upon it. And on the table there were little figures like chessmen who went to and fro doing this and that. And I knew that each chessman was the idolum or puppet representative of some of the great presences that stood by. And the acts and motions of each chessman were a moving portrait, a mimicry or pantomime, which delineated the inmost nature of his giant master. And these chessmen are men and women as they appear to themselves and to one another in this world. And the silver table is Time. And those who stand and watch are the immortal souls of those same men and women. Then vertigo and terror seized me and, clutching at my Teacher, I said, 'Is *that* the truth? Then is all that I have been seeing in this country false? These conversations between the Spirits and the Ghosts—were they only the mimicry of choices that had really been made long ago?'

'Or might ye not as well say, anticipations of a choice to be made at the end of all things? But ye'd do better to say neither. Ye saw the choices a bit more clearly than ye could see them on Earth: the lens was clearer. But it was still seen through the lens. Do not ask of a vision in a dream more than a vision in a dream can give.'

'A dream? Then—then—am I not really here, Sir?'

'No, Son,' said he kindly, taking my hand in his. 'It is not so good as that. The bitter drink of death is still before you. Ye are only dreaming. And if ye come to tell of what ye have seen, make it plain that it was but a dream. See ye make it very plain. Give no poor fool the pretext to think ye are claiming knowledge of what no mortal knows. I'll have no Swedenborgs and no Vale Owens among my children.'

'God forbid, Sir,' said I, trying to look very wise.

'He *has* forbidden it. That's what I'm telling ye.' As he said this he looked more Scotch than ever. I was gazing steadfastly on his face. The vision of the chessmen had faded, and once more the quiet woods in the cool light before sunrise were about us. Then, still looking at his face, I saw there something that sent a quiver through my whole body. I stood at that moment with my back to the East and the mountains, and he, facing me, looked towards them. His face flushed with a new light. A fern, thirty yards behind him, turned golden. The eastern side of every tree-trunk grew bright. Shadows deepened. All the time there had been bird noises, trillings, chatterings, and the like; but now suddenly the full chorus was poured from every branch; cocks were crowing, there was music of hounds, and horns; above all this ten thousand tongues of men and woodland angels and the wood itself sang. 'It comes, it comes!' they sang. 'Sleepers awake! It comes, it comes, it comes.' One dreadful glance over my shoulder I essayed—not long enough to see (or did I see?) the rim of the sunrise that shoots Time dead with golden arrows and puts to flight all phantasmal shapes. Screaming, I buried my face in the fold of my Teacher's robe. 'The morning! The morning!' I cried, 'I am caught by the morning and I am a ghost.' But it was too late. The light, like solid blocks, intolerable of edge and weight, came thundering upon my head. Next moment the folds of my Teacher's garment were only the folds of the old ink-stained cloth on my study table which I had pulled down with me as I fell from my chair. The blocks of light were only the books which I had pulled off with it, falling about my head. I awoke in a cold room, hunched on the floor beside a black and empty grate, the clock striking three, and the siren howling overhead.